Recent Titles in This Series

156 V. N. Gerasimov, N. G. Nesterenko, and A. I. Valitskas, Three Papers on Algebras and Their Representations
155 O. A. Ladyzhenskaya and A. M. Vershik, Editors, Proceedings of the St. Petersburg Mathematical Society, Volume I
154 V. A. Artamonov, et al., Selected Papers in K-Theory
153 S. G. Gindikin, Editor, Singularity Theory and Some Problems of Functional Analysis
152 H. Draškovičová, et al., Ordered Sets and Lattices II
151 I. A. Aleksandrov, L. A. Bokut', and Yu. G. Reshetnyak, Editors, Second Siberian Winter School "Algebra and Analysis"
150 S. G. Gindikin, Editor, Spectral Theory of Operators
149 V. S. Afraĭmovich, et al., Thirteen Papers in Algebra, Functional Analysis, Topology, and Probability, Translated from the Russian
148 A. D. Aleksandrov, O. V. Belegradek, L. A. Bokut', and Yu. L. Ershov, Editors, First Siberian Winter School in Algebra and Analysis
147 I. G. Bashmakova, et al., Nine Papers from the International Congress of Mathematicians 1986
146 L. A. Aĭzenberg, et al., Fifteen Papers in Complex Analysis
145 S. G. Dalalyan, et al., Eight Papers Translated from the Russian
144 S. D. Berman, et al., Thirteen Papers Translated from the Russian
143 V. A. Belonogov, et al., Eight Papers Translated from the Russian
142 M. B. Abalovich, et al., Ten Papers Translated from the Russian
141 Kh. Drashkovicheva, et al., Ordered Sets and Lattices
140 V. I. Bernik, et al., Eleven Papers Translated from the Russian
139 A. Ya. Aĭzenshtat, et al., Nineteen Papers on Algebraic Semigroups
138 I. V. Kovalishina and V. P. Potapov, Seven Papers Translated from the Russian
137 V. I. Arnol'd, et al., Fourteen Papers Translated from the Russian
136 L. A. Aksent'ev, et al., Fourteen Papers Translated from the Russian
135 S. N. Artemov, et al., Six Papers in Logic
134 A. Ya. Aĭzenshtat, et al., Fourteen Papers Translated from the Russian
133 R. R. Suncheleev, et al., Thirteen Papers in Analysis
132 I. G. Dmitriev, et al., Thirteen Papers in Algebra
131 V. A. Zmorovich, et al., Ten Papers in Analysis
130 M. M. Lavrent'ev, et al., One-dimensional Inverse Problems of Mathematical Physics
129 S. Ya. Khavinson; translated by D. Khavinson, Two Papers on Extremal Problems in Complex Analysis
128 I. K. Zhuk, et al., Thirteen Papers in Algebra and Number Theory
127 P. L. Shabalin, et al., Eleven Papers in Analysis
126 S. A. Akhmedov, et al., Eleven Papers on Differential Equations
125 D. V. Anosov, et al., Seven Papers in Applied Mathematics
124 B. P. Allakhverdiev, et al., Fifteen Papers on Functional Analysis
123 V. G. Maz'ya, et al., Elliptic Boundary Value Problems
122 N. U. Arakelyan, et al., Ten Papers on Complex Analysis
121 D. L. Johnson, The Kourovka Notebook: Unsolved Problems in Group Theory
120 M. G. Kreĭn and V. A. Jakubovič, Four Papers on Ordinary Differential Equations
119 V. A. Dem'janenko, et al., Twelve Papers in Algebra
118 Ju. V. Egorov, et al., Sixteen Papers on Differential Equations

(*Continued in the back of this publication*)

Three Papers on Algebras and Their Representations

American Mathematical Society

TRANSLATIONS

Series 2 • Volume 156

Three Papers on Algebras and Their Representations

V. N. Gerasimov
N. G. Nesterenko
A. I. Valitskas

American Mathematical Society
Providence, Rhode Island

Translated by MIRA BERNSTEIN and BORIS M. SCHEIN
from original Russian manuscripts
Translation edited by SIMEON IVANOV

1991 *Mathematics Subject Classification.* Primary 16Gxx, 16Nxx.

Library of Congress Cataloging-in-Publication Data
Gerasimov, V. N.
　Three papers on algebras and their representations / V. N. Gerasimov, N. G. Nesterenko, A. I. Valitskas.
　　p. cm. – (American Mathematical Society translations, ISSN 0065-9290; ser. 2, v. 156)
　　Includes bibliographical references.
　　ISBN 0-8218-7503-5
　　1. Representations of algebras. 2. Associative algebras. 3. Radical theory. I. Nesterenko, N. G. II. Valitskas, A. I. III. Title. IV. Title: 3 papers on algebras and their representations. V. Series.
QA3.A572 ser. 2, vol. 156
[QA251.5]
510 s–dc20　　　　　　　　　　　　　　　　　　　　　　　　　　　　　　　　　93-20884
[512$'$.4]　　　　　　　　　　　　　　　　　　　　　　　　　　　　　　　　　　　　CIP

Copying and reprinting. Individual readers of this publication, and nonprofit libraries acting for them, are permitted to make fair use of the material, such as to copy an article for use in teaching or research. Permission is granted to quote brief passages from this publication in reviews, provided the customary acknowledgment of the source is given.
　Republication, systematic copying, or multiple reproduction of any material in this publication (including abstracts) is permitted only under license from the American Mathematical Society. Requests for such permission should be addressed to the Manager of Editorial Services, American Mathematical Society, P.O. Box 6248, Providence, Rhode Island 02940-6248. Requests can also be made by e-mail to reprint-permission@math.ams.org.
　The owner consents to copying beyond that permitted by Sections 107 or 108 of the U.S. Copyright Law, provided that a fee of $1.00 plus $.25 per page for each copy be paid directly to the Copyright Clearance Center, Inc., 222 Rosewood Drive, Danvers, Massachusetts 01923. When paying this fee please use the code 0065-9290/93 to refer to this publication. This consent does not extend to other kinds of copying, such as copying for general distribution, for advertising or promotional purposes, for creating new collective works, or for resale.

Copyright ©1993 by the American Mathematical Society. All rights reserved.
Translation authorized by the
All-Union Agency for Author's Rights, Moscow.
Printed in the United States of America.
The American Mathematical Society retains all rights
except those granted to the United States Government.
The paper used in this book is acid-free and falls within the guidelines
established to ensure permanence and durability. ∞
♻ Printed on recycled paper.

This publication was typeset using $\mathcal{A}_{\mathcal{M}}\mathcal{S}$-TEX,
the American Mathematical Society's TEX macro system.
10 9 8 7 6 5 4 3 2 1　　97 96 95 94 93

Contents

Foreword ... ix

Part I: Free Associative Algebras and Inverting Homomorphisms of Rings
    by V. N. Gerasimov ... 1

Introduction ... 3

Chapter 1: Free Algebras and Algebras with a Single Relation ... 9
    §1. Some corollaries of Theorem 1 ... 10
    §2. Reduction of Theorem 1 to a problem about subspaces of the field of rational functions ... 14
    §3. Distributive sublattices of modular lattices ... 17
    §4. Basic rings and basic modules ... 20
    §5. Corollaries of Lemma 4.3 ... 30
    §6. Special sets and special inference ... 36
    §7. Proof of the Main Lemma ... 41

Chapter 2: Inverting Homomorphisms of Rings ... 51
    §1. Description of the construction of localization ... 52
    §2. Localizations by independent sets of matrices ... 67

References ... 75

Part II: Representations of Algebras by Triangular Matrices
    by N. G. Nesterenko ... 77

Conventions and Notation ... 79

Introduction ... 81

Chapter 1: Representability of Triangular Categories and Graded Algebras ... 87
    §1. Preliminary remarks on representable triangular categories ... 87
    §2. Quasi-identities of representable triangular categories ... 90
    §3. Representations of graded algebras ... 94

Chapter 2: Representation of Algebras by Triangular Matrices.
Algebras with Diagonal 99
§1. Statement of the basic theorem and its corollaries 99
§2. Algebras with diagonal, adjoining a diagonal 103
§3. Proof of the basic theorem 107
§4. Representation of algebras with diagonal 111
§5. Certain counterexamples 114

Chapter 3: Special Representations of Nilpotent Graded Algebras 117
§1. Embedding connected graded algebras 117
§2. Certain applications 121

References 124

Part III: Embedding Rings in Radical Rings and Rational Identities of Radical Algebras
by A. I. Valitskas 125

Introduction 127
§1. Building universal constructions 128
§2. Investigation of complexity of quasivarieties of algebraic systems 129
§3. Investigation of when concrete classes of algebraic systems belong to a given quasivariety 131

Index of Notation 137

Chapter 1: Absence of a Finite Basis of Quasi-identities for the Quasivariety of Rings Embeddable in Radical Rings 139
§1. A construction of matrix localization suggested by V. N. Gerasimov 139
§2. Homomorphisms into radical rings 142
§3. R-independent sets of matrices 149
§4. Examples of A_n-rings 155

Chapter 2: Examples of Noninvertible Rings Embeddable in Groups 157
§1. The structure of the semigroup $\overline{R^*}$ 158
§2. Embedding the semigroup $\overline{R^*}$ in a group 168

Chapter 3: Representation of Finite-Dimensional Lie Algebras in Radical Rings 177

Chapter 4: Rational Identities of Radical Algebras 183
§1. Rational identities of a special form 184
§2. Proof of Theorem 4.1 by means of Theorem 4.2 185
§3. Proof of Theorem 4.2 188

References 194

Foreword

I am pleased to write the foreword to this book not only because it contains doctoral dissertations of my former students, V. N. Gerasimov, A. I. Valitskas, and N. G. Nesterenko, but also because I really think that these dissertations deserve to be published in such an important series as Translations of the American Mathematical Society.

The dissertation of V. N. Gerasimov is a part of the theory of noncommutative matrix localizations discovered by P. M. Cohn. Inspired by some ideas of Gabriel and Zisman on localizations of categories, Gerasimov found a construction of matrix localization that is not directly related to (prime) matrix ideals of P. M. Cohn but rather deals with localizations of arbitrary subsets of matrices over a ring. In particular, this construction makes it possible to localize an arbitrary subset of nonzero elements of a ring without imposing the Ore condition. The idea of this approach was that after appropriate extension of the class of "objects" there are no localizations other than Ore localizations (kind of a paradox!). Independently but later these ideas were rediscovered by P. Malcolmson. The results of Gerasimov and Malcolmson improved our understanding of Cohn's theory.

A. I. Valitskas applied ideas and constructions of Gerasimov to embeddings of rings into radical rings in the sense of N. Jacobson. As a result, he developed a theory that was essentially parallel to Cohn's theory of embeddings of rings into skew fields.

N. G. Nesterenko solved some important problems of A. Z. Anan′in and Bergman about representations of (infinite-dimensional) algebras and categories in (triangular) matrices over commutative rings. The series of works on such embeddings was initiated by A. I. Mal′cev in the 1940s and then continued by S. A. Amitsur in the 1950s. The latter work put the embedding problem in the context of PI-theory and this is how it was considered by all subsequent authors including N. G. Nesterenko.

V. N. Gerasimov, A. I. Valitskas, and N. G. Nesterenko did much of their work while participating in my seminar, which was organized as a "daughter" seminar of the seminar of A. I. Shirshov (the latter was a "daughter" seminar of the seminar of A. I. Mal′cev). It is noteworthy that participants of these seminars worked with (and learned from) several generations of mathematicians. For example, Gerasimov was the "coadvisor" of Valitskas, while Nesterenko was "coadvised" by I. V. L'vov and A. Z. Anan'in. This shows the role of "seminars" in the Russian mathematical school.

L. A. Bokut'

March 1993

Part I

Free Associative Algebras and Inverting Homomorphisms of Rings

V. N. Gerasimov

Introduction

The important role in ring theory, as in any other algebraic theory, belongs to the study of the universal constructions, or using the language of category theory, functors that have a right adjoint. Examples of such functors are a ring of matrices over a given ring, a quotient field of a commutative integral domain, a free product of algebras over a field, and so on.

The majority of universal constructions can be described in terms of generators and relations, although other, more convenient descriptions are used more frequently.

This thesis is a study of the following two universal constructions:

1. free associative algebras and algebras with a single relation;
2. localizations of associative rings, i.e., rings obtained by formal adjunction of the inverses of given elements to a ring with unity.

One of the universal methods for studying algebras defined by relations is the composition method, suggested by A. I. Shirshov [15] for Lie algebras and applied by L. A. Bokut' [4, 5] to associative algebras.

The essence of this method is an algorithm for extending a given set of defining relations to an equivalent set of relations $\{f_\alpha = 0 \mid \alpha \in A\}$ such that in the ideal $\mathrm{Idl}(f_\alpha \mid \alpha \in A)$, generated by polynomials f_α in a free algebra, the leading words of all elements belong to the ideal $\mathrm{Idl}(\overline{f_\alpha} \mid \alpha \in A)$ generated by the leading words $\overline{f_\alpha}$ of polynomials f_α. This algorithm is called the closure with respect to composition (see [4, 5]).

In simple cases the composition method is often used implicitly. An example of such implicit usage is any standard proof of the Birkhoff-Witt Theorem.

Unfortunately, the above algorithm of closure with respect to composition often yields a huge system of relations, even for a single initial relation.

Another method is related to generating series and so far has been applied mostly to homogeneous algebras.

Let $A = \bigoplus_{n \geq 0} A^n$ be a graded algebra, with A^n, $n \geq 0$, being subspaces of homogeneous elements of degree n. The formal power series

$$H_A(t) \rightleftharpoons \sum_{n \geq 0} (\dim A^n) t^n$$

is called the Hilbert-Poincaré series of A.

Using these series, E. S. Golod [8] proved the well-known theorem on the lower estimate for dimensions of the homogeneous components in terms of the degrees of homogeneous relations. This theorem has a number of interesting corollaries.

The following important problem in the theory of homogeneous algebras originating from homological algebra (see Kostrikin–Shafarevitch [10], Govorov [8]) remained unsolved until recently.

Let A be a graded associative algebra with a finite number of relations. Is the series $H_A(t)$ a rational function?

Recently V. A. Ufnarovskiĭ has constructed a 2-generator graded algebra with two homogeneous relations such that its Hilbert-Poincaré series is not rational.

For algebras with a single relation the rationality of the Hilbert-Poincaré series is an easy consequence of the results of this thesis (see [20]).

We also want to mention two yet unsolved problems.

1. Is the word problem solvable for algebras with a single relation? The answer is unknown even for semigroups, although in this case there is significant progress (see S. I. Adyan [1]).

2. Is the freeness theorem true for algebras with a single relation? Note the result of J. and T. Lewin [27], who proved the freeness theorem for algebras without zero divisors defined by one homogeneous relation.

In Chapter 1 we study algebras with a single relation, or equivalently, 1-generator ideals of a free associative algebra over a field.

Let k be an arbitrary field, let $\langle X \rangle$ be the semigroup with unity freely generated by X, let d be a positive integer-valued degree function on $\langle X \rangle$, let $F = k\langle X \rangle$ be the free associative algebra, and let F^n be the subspace generated by all monomials of degree $n \geq 0$. Elements of F^n are called *homogeneous* (or d-homogeneous, if the degree function d is not fixed).

Let U be a linear space, $\mathscr{V}$ a set of its subspaces. A basis $\mathscr{B}$ of the space U is called a *common basis* for the set $\mathscr{V}$ if for any subspace $V \in \mathscr{V}$ the set $\mathscr{B} \cap V$ is a basis of V.

The main result of Chapter 1 is the following theorem.

THEOREM 1. *Let f be a homogeneous element of F. Then the set of subspaces $\{F^m f F^n \mid m, n \geq 0\}$ of F has a common basis.*

Using Theorem 1, Backelin [20] deduced the rationality of the Hilbert-Poincaré series for any algebra with a single relation and obtained a number of other corollaries.

REFORMULATION OF THEOREM 1. *For any homogeneous element $f \in F$ the set of subspaces $\{F^m f F^n \mid m, n \geq 0\}$ generates a distributive sublattice in the lattice of all subspaces of F.*

Denote by h_d the highest d-homogeneous component of an element $h \in F$ and by H_d the set $\{h_d \mid h \in H\}$ of d-homogeneous components of all elements of a subset $H \subset F$ (by definition, we set $0_d \rightleftharpoons 0$).

The following analog of the composition lemma (Proposition 4.1) is obtained in the thesis as a corollary of Theorem 1.

If an element $f \in F$ satisfies the condition

$$f_d F_d \cap F_d f_d = (fF \cap Ff)_d, \tag{1}$$

then the highest component of any element from the ideal $\mathrm{Idl}(f)$ belongs to the ideal $\mathrm{Idl}(f_d)$ generated by the highest component of f.

Condition (1) is an analog of closeness with respect to composition for the one-element set $\{f\}$. The difference is that instead of the leading words we consider highest homogeneous components. Clearly, in this case there are far fewer compositions.

This corollary immediately implies that the word problem is algorithmically solvable for any algebra defined by the relation $f = 0$, where f satisfies the condition (1) (Theorem 2).

In particular, condition (1) is satisfied if the highest component f_d of f does not have proper two-sided divisors, that is, if any element that is both a right and left divisor of f_d is either invertible or is proportional to f_d.

In addition, we also prove the freeness theorem for a particular class of algebras given by a single relation (Theorem 3).

Chapter 2 is dedicated to another universal construction, namely localization.

We recall a few definitions. A homomorphism $f: R \to R_1$ of rings with unity is called S-*inverting* if it maps all elements from a set $S \subset R$ into invertible elements of R_1. Similarly, one defines Σ-*inverting* homomorphisms for a set Σ of rectangular matrices over a ring R. The second notion is clearly a generalization of the first one. Fix a ring R and a set Σ; then among all Σ-inverting homomorphisms there exists a unique homomorphism $n_\Sigma: R \to R\Sigma^{-1}$ universal in the sense that for any Σ-inverting homomorphism $f: R \to R_1$ there exists a unique homomorphism $h: R\Sigma^{-1} \to R_1$ such that the following diagram is commutative.

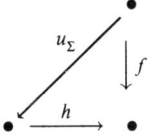

This universal Σ-inverting homomorphism is called the *localization of the ring R by the set Σ*. To construct it we can, for instance, formally adjoin to R all the elements of matrices inverse to matrices from Σ.

Localization plays an important role in commutative algebra, where a universal S-inverting ring is constructed as the ring of formal fractions with denominators belonging to a set S.

For noncommutative rings such a general construction does not exist so far. The very important Ore theorem [29], [30] outlines boundaries within which it is still possible to apply the construction of fractions. Apart from this theorem, until recently there existed only very limited methods for inverting elements of noncommutative rings.

Note that the study of localization can be carried out using the method of inverting and defining relations. Using this method A. I. Mal′cev [11]–[13] solved the problem posed by van der Waerden and found necessary and sufficient conditions for a semigroup to be embeddable into a group. The most prominent success in the field was the negative answer in 1966 to the problem posed by Mal′cev about whether Mal′cev's conditions are sufficient for a ring to be embeddable into a skew field (L. A Bokut′ [2], [3], Bowtell [23], Klein [26]).

Significant progress in the study of localizations was made by Cohn in the beginning of the 1970s. His book [9] *Free rings and their relations* became a landmark in the development of this theory. Cohn discovered the following two circumstances:

1. the analysis of localizations is directly related to the theory of near free rings;

2. the analysis of localizations becomes much simpler if one considers not only inverses of elements, but also inverses in the set of matrices.

(Note that Bergman [22] went even further and suggested inverting homomorphisms of finitely-generated projective modules, but his idea has not yet been developed.)

Cohn introduced and analyzed important classes of near free rings: FI-rings, rings with weak algorithm, and a number of other classes. These rings and their generalizations were later studied by many authors (Bergman, Bowtell, Dicks, Fisher, Klein, Lewin, Skornyakov, and others). Cohn used FI-rings in his proof of the existence of amalgamated products of skew fields [25]. In 1972 Cohn described all homomorphisms of rings into skew fields and found a necessary and sufficient condition (in the form of a system of quasi-identities) for an integral domain to be embeddable into a skew field.

In Chapter 2 we describe a construction of localization of an arbitrary ring R by an arbitrary set Σ of rectangular matrices. It turns out that after adding to the set Σ certain matrices that do not affect localization, the construction becomes not much harder than the construction of Ore's fractions. Theorem 4 contains the most essential statements concerning the construction, in particular, the criterion for equality of two elements belonging to a universal Σ-inverting ring, which is similar to the criterion for equality of two fractions in Ore's construction. Both Ore's theorem and Cohn's theorems (on embeddability of rings into skew fields) can be easily deduced from Theorem 4.

A set Σ is called *potentially invertible* if all matrices from Σ are invertible in an extension of the original ring, or equivalently, if the universal

Σ-inverting homomorphism is injective. A ring is called *invertible* if the set of all its nonzero elements is potentially invertible.

Using Theorem 4, we describe the kernel of the universal Σ-inverting homomorphism and produce a system of quasi-identities that are satisfied if and only if a given set is potentially invertible (Theorem 5).

In the aforementioned paper [25] Cohn posed the question of whether the multiplicative semigroup of nonzero elements of any 2- FI-ring is embeddable into a group.

This question is answered in the positive by the following theorem.

THEOREM 7. *Any 2- FI-ring is invertible.*

Let $\mathscr{K}_0$ be the class of integral domains, $\mathscr{K}_1$ the class of rings with the multiplicative semigroup of nonzero elements embeddable into a group, $\mathscr{K}_2$ the class of invertible rings, and $\mathscr{K}_3$ the class of rings embeddable into skew fields. The inclusions $\mathscr{K}_0 \supseteq \mathscr{K}_1 \supseteq \mathscr{K}_2 \supseteq \mathscr{K}_3$ are obvious. In 1937 Mal'cev proved that $\mathscr{K}_0 \neq \mathscr{K}_1$. The example of Bokut' [2], [3] which solves Mal'cev's problem (whether or not $\mathscr{K}_1 = \mathscr{K}_3$) shows that $\mathscr{K}_1 \neq \mathscr{K}_2$. Since there are many examples of 2- FI-rings nonembeddable into skew fields (the first such example was constructed by Cohn [24]), Theorem 7 implies that $\mathscr{K}_2 \neq \mathscr{K}_3$.

Theorem 8 provides a sufficient condition for a ring to be an n- FI-ring. This theorem gives an answer to the question of Bergman [22] about the interconnection of dependency parameters for rings R and $R\Sigma^{-1}$, where Σ is the set of all square matrices of a fixed order $l > 0$. It turns out that if R is an n- FI-ring, then $R\Sigma^{-1}$ is an $(n - 2l)$- FI-ring (Theorem 9). In particular, if $S = R\setminus\{0\}$ is the set of all nonzero elements of an n- FI-ring R, then $R\Sigma^{-1}$ is an $(n - 2)$- FI-ring.

The thesis consists of two chapters. The numbering of formulas, lemmas, and paragraphs is separate in each chapter, the numbering of theorems is continuous throughout the thesis. Since there are no cross references between the chapters, we do not have to indicate chapter numbers in the references.

The results of the thesis were presented at the seminars "Algebra and Logics", "Ring Theory", "Associative Rings and Lie Rings" in the Institute of Mathematics, Siberian Section of the Academy of Sciences of the USSR and in Novosibirsk State University, at the seminars in Moscow and Omsk State Universities, at the 3rd All-Union Conference on Mathematical Logics (Novosibirsk, 1974) and the 4th All-Union Algebraic Conference (Krasnoyarsk, 1979).

The author wants to express his sincerest gratitude to his advisor, professor L. A. Bokut', for scientific guidance and to the corresponding member of the Academy of Sciences of the USSR, A. I. Shirshov, for his support and concern.

Chapter 1
Free Algebras and Algebras with a Single Relation

Let k be an arbitrary field, let F be a free associative algebra with unit over k, and let X be a set of free generators for algebra F. We shall consider only integer-valued *degree functions* d satisfying the condition

$$d\left(\sum_{m\in\langle X\rangle} \lambda_m m\right) = \max_{\lambda_m \neq 0} d(m), \quad \lambda_m \in k. \tag{1}$$

Each such degree function defines a grading $F = \bigoplus_{n\geq 0} F_d^n$, where F is a subspace generated by all the monomials of degree $n \geq 0$. If $d(x) > 0$ for any $x \in X$, a degree function is called *positive*. The subscript d will be omitted if it is clear from the context which degree function is being referred to.

Fix a positive function d. Elements of subspaces F_d^n, $n \geq 0$, will be called *homogeneous elements*.

The main result of Chapter 1 is the following.

THEOREM 1. *Let f be a homogeneous element of F. Then the set of subspaces*

$$\{F^m f F^n \mid m, n \geq 0\}$$

has a common basis.

It will be proved below (Lemma 2.1) that the following statement is equivalent to Theorem 1.

THEOREM 1'. *Let f be a homogeneous element of F, and let m be a natural number. Then the set*

$$\{F^{m_1} f F^{m_2} \mid m_1 + m_2 = m, \ m_1, m_2 \geq 0\}$$

generates a distributive sublattice in a lattice of subspaces of the space $F^{m+d(f)}$.

§1. Some corollaries of Theorem 1

LEMMA 1.1. *Let f be a homogeneous element of algebra F. Then*

$$F^m f F^{k-m} \cap F^n f F^{k-n} = F^m (f F^{n-m} \cap F^{n-m} f) F^{k-n},$$

where $0 \leq m \leq n \leq k$, m, n, k are natural numbers.

PROOF. Since subspaces $F^{m_1} \cdot F^{m_2}$ and $f^{m_1} \otimes F^{m_2}$ are isomorphic for any natural m_1, m_2, we can use the following well-known property of tensor products.

Let A_1, A_2 be subspaces of A. Then

$$(A_1 \otimes B) \cap (A_2 \otimes B) = (A_1 \cap A_2) \otimes B.$$

Since $F^{m_1} f F^{m_2} \subseteq F^{m_1} F^{d(f)} F^{m_2}$, it follows that

$$F^m f F^{k-m} \cap F^n f F^{k-n} = F^m f F^{n-m} F^{k-n} \cap F^m F^{n-m} f F^{k-n}$$
$$= F^m (f F^{n-m} \cap F^{n-m} f) F^{k-n}.$$

The lemma is proved.

LEMMA 1.2. *Let $\{u_\lambda \in F^m \mid \lambda \in \Lambda\}$, $\{v_\lambda \in F^n \mid \lambda \in \Lambda\}$ be finite families, and let a_1, a_2 be nonzero elements of F. Then*

$$\sum_\lambda u_\lambda a_1 v_\lambda = 0 \Leftrightarrow \sum_\lambda u_\lambda a_2 v_\lambda = 0.$$

PROOF. It suffices to note that the homomorphism of spaces

$$\sum_\lambda u_\lambda \otimes v_\lambda \to \sum_\lambda u_\lambda a v_\lambda$$

is an isomorphism.

Theorem 1 enables us to study 1-generator ideals of a free algebra that are generated by an element f subject to the following condition:

$$f_d F_d \cap F_d f_d = (fF \cap Ff)_d, \tag{2}$$

where the notation is the same as in the Introduction.

Consider the equation

$$T(f) = g, \tag{3}$$

where $f, g \in F$, T is an operator belonging to the algebra of multiplications, i.e.,

$$T(y) = \sum u_\lambda y v_\lambda, \qquad u_\lambda, v_\lambda \in F, \tag{4}$$

y is a variable not contained in the set X of free generators. We shall consider (3) as an equation for the unknown operator T, i.e., for elements u_λ, v_λ.

PROPOSITION 1.3. *Let $f \in F$ be an element satisfying (2), let g be an arbitrary element of F, and let T be an operator from the algebra of multi-*

plication. Then the following statements are equivalent:
1) $T(f) = g$;
2) *operator T has the form*

$$T(y) = T_0(y) + \sum_\nu T_\nu(yp_\nu - q_\nu y),$$

where

$$T_0(y) = \sum_\mu u_\mu y v_\mu, \qquad d(u_\mu) + d(v_\mu) \leq d(g) - d(f),$$

T_ν *are arbitrary operators*, $p_\nu, q_\nu \in F$, $fp_\nu = q_\nu f$.

PROOF. The element $T(y)$ can be uniquely presented in the following form: $T(y) = \sum_{\nu \in N(T)} T_\nu(y)$, where $N(T)$ is a finite set of pairs $\nu = (m, n)$ of nonnegative integers, and the operators $T_\nu = T_{(m,n)}$ have the form

$$T_\nu(y) = \sum_\mu u_\mu y v_\mu, \qquad u_\mu \in F^m, \ v_\mu \in F^n \tag{5}$$

and are not identically zero. We call operators of the form (5) *homogeneous*, and numbers m, n *right* and *left degrees* of the operator, respectively. Let

$$N_s(T) \rightleftharpoons \{(m, n) \in N(T) \mid m + n = s\}, \qquad s \geq 0;$$
$$s(T) \rightleftharpoons \max\{s \mid N_s(T) \neq \varnothing\}.$$

We prove that if $s(T) > d(g) - d(f)$, then T can be represented in the form

$$T(y) = T_1(y) + \sum_\nu T_\nu(yp_\nu - q_\nu y), \qquad fp_\nu = q_\nu f, \tag{6}$$

where either $s(T_1) < s(T)$, or $s(T_1) = s(T) \rightleftharpoons s$, $|N_s(T_1)| < |N_s(T)|$. (From now on, the cardinality of a set A is denoted by $|A|$.) The desired statement will then follow by induction.

Let $s \rightleftharpoons s(T) > d(g) - d(f)$. By equating homogeneous components of degree $s + d(f)$ in equation (3), we obtain $\sum_{\nu \in N_s(T)} T_\nu(f) = 0$. Let $\nu_0 = (m_0, n_0)$ be a pair from the set $N_s(T)$ for which the value of m is minimal. If $N_s(T) = \{\nu_0\}$, then by Lemma 1.2 the operator $T_{\nu_0}(y)$ is identically zero, which is impossible by assumption. Let $N'_s \rightleftharpoons N_s(T) \setminus \{\nu_0\}$. Using Theorem 1, Lemma 1.1, and condition 2, we obtain the following result:

$$T_{\nu_0}(f_d) = - \sum_{(m,n) \in N'_s} T_{(m,n)}(f_d) \in F^{m_0} f_d F^{n_0} \cap \sum_{(m,n) \in N'_s} F^m f_d F^n$$
$$\subseteq \sum_{(m,n) \in N'_s} F^{m_0}(fF \cap Ff)_d F^n.$$

Hence,

$$T_{\nu_0}(f_d) = \sum_{(m,n) \in N'_s} T'_{(m,n)}((w_{m,n})d),$$

where $T'_{m,n}$ are homogeneous operators, with left and right degrees equal to m and n, respectively, $w_{m,n} = fp_{m,n} = q_{m,n}f$, $d(w_{m,n}) = d(f) + m - m_0$ for some $p_{m,n}, q_{m,n} \in F$. By Lemma 1.2, we get

$$T_{\nu_0}(y) = \sum_{(m,n) \in N'_s} T'_{m,n}(y(p_{m,n})_d);$$

$$T(y) = \sum_{\nu \in N(T) \setminus \{\nu_0\}} T_\nu(y) + \sum_{(m,n) \in N'_s} (T'_{m,n}(yp_{m,n} - q_{m,n}y) + T'_{m,n}(q_{m,n}y)$$

$$+ T'_{m,n}(y((p_{m,n})_d - p_{m,n}))).$$

Setting

$$T_1(y) \rightleftharpoons \sum_{\nu \in N(T) \setminus \{\nu_0\}} T_\nu(y) + \sum_{(m,n) \in N'_s} T'_{m,n}(y((p_{m,n})_d - p_{m,n}) + q_{m,n}y),$$

we obtain the desired decomposition (6).

The proposition is proved.

PROPOSITION 1.4. *Let $f \in F$ be an element satisfying (2). Then the maximum-degree homogeneous component of any element from the ideal* $\mathrm{Idl}(f)$ *generated by f is contained in the ideal* $\mathrm{Idl}(f_d)$ *generated by f_d.*

PROOF. By proposition 1.3, g can be presented in the form

$$g = \sum_\mu u_\mu f v_\mu, \qquad d(u_\mu) + d(f) + d(v_\mu) \le d(g). \tag{7}$$

By equating the maximal-degree components, we obtain the desired statement.

THEOREM 2. *Let $f \in F$ be an element satisfying the condition*

$$f_d F_d \cap F_d f_d = (fF \cap Ff)_d.$$

Then

1) *for the algebra $\langle X; f = 0 \rangle$ the equality problem is algorithmically decidable (if the base field is k constructible);*

2) *an ideal $\mathrm{Idl}(f)$, viewed as an (F, F)-bimodule, is defined by a system of relations of the form $fp = qf$, where $p, q \in F$.*

PROOF. Verification of (7) for a given element g can be reduced to the problem of compatibility of a certain finite system of linear equations. This yields the first statement. To prove the second statement, it suffices to set $g = 0$ in Proposition 1.3.

The theorem is proved.

We call an element $h \in F$ a *proper two-sided divisor* of an element $f \in F$, $0 < d(h) < d(f)$, if $f = hf' = f''h$ for some $f', f'' \in F$.

LEMMA 1.5. *Let $f \in F$, and let f_d have no proper two-sided divisors. Then*

1) *f satisfies (2);*

2) *if* $fp = qf$ *for* $p, q \in F$, *then* $p = rf + \alpha$, $q = fr + \alpha$ *for some* $r \in F$, $\alpha \in k \cdot 1$.

PROOF. 1) Since the map $h \to h_d$ is a homomorphism of multiplicative semigroups, the inclusion $(fF \cap Ff)_d \subseteq f_d F_d \cap F_d f_d$ is obvious. Let $w = f_d u = v f_d \in f_d F^m \cap F^m f_d$, where $u, v \in F^m$. The following is a well-known property of the multiplicative semigroup of homogeneous elements in a free algebra (see, for instance, [9, Proposition 6.6.3]): if $ab = cd$, then there exists a homogeneous element h such that either $a = ch$, $d = hb$, or $c = ah$, $b = hd$.

Applying this property to the equation $f_d u = v f_d$ we see that for a homogeneous element h either $f_d = vh = hu$, or $u = h f_d$, $v = f_d h$. We can assume the latter, since in case of the former, h is a two-sided proper divisor of f_d if $d(h) \neq 0$. Setting $u_1 \rightleftharpoons hf$, $v_1 \rightleftharpoons fh$, we obtain $w = (fu_1)_d = (v_1 f)_d \in (fF \cap Ff)_d$.

To prove 2), we use induction on the number $d(p) = d(q)$. If this number equals zero the statement is obvious. Let $d(p) = d(q) > 0$, $fp = qf$. Then $f_d p_d = q_d f_d$ and similarly to the proof of 1) above, $p_d = h f_d$, $q_d = f_d h$ for some homogeneous element h. Set $p' \rightleftharpoons p - hf$, $q' \rightleftharpoons q - fh$. Then $fp' = q'f$, $d(p') < d(p)$. By the induction assumption $p' = r'f + \alpha$, $q' = fr' + \alpha$ for some $r' \in F$, $\alpha \in k \cdot 1$. Setting $r \rightleftharpoons r' + h$, we arrive at $p = rf + \alpha$, $q = fr + \alpha$.

The lemma is proved.

COROLLARY 1.6. *If the maximal-degree homogeneous component* f_d *of* $f \in F$ *has no proper two-sided divisors, then*

1) *the equality problem is algorithmically solvable for the algebra* $\langle X; f = 0 \rangle$;

2) *the ideal* $\mathrm{Idl}(f)$ *is "free as a two-sided ideal". This means that if* $T(f) = 0$, *where* $T(y)$ *is an operator belonging to the algebra of multiplications, then* $T(y) = S(f, y) - S(y, f)$, *where* $S(y_1, y_2)$ *is an operator depending on two variables and having the form*

$$S(y_1, y_2) = \sum_\nu u_\nu y_1 v_\nu y_2 w_\nu, \qquad u_\nu, v_\nu, w_\nu \in F.$$

Both statements follow immediately from Theorem 2 and Lemma 1.5.

THEOREM 3. *Let* f *be an element of algebra* $F = k\langle X \rangle$, *let* x_0 *be an element of the set* X *of free generators, let* $f \in k\langle X \setminus \{x_0\} \rangle$, *and let* f_{x_0} *be the leading term of* f *with respect to* x_0. *If* $(f_{x_0})_d$ *has no proper two-sided divisors, then*

$$k\langle X \setminus \{x_0\} \rangle \cap \mathrm{Idl}(f) = \{0\},$$

i.e., the set $X \setminus \{x\}$ *generates a free subalgebra in the algebra* $\langle X; f = 0 \rangle$.

PROOF. Let $g \in k\langle X\setminus\{x_0\}\rangle \cap \mathrm{Idl}(f)$. Define a new degree function:

$$d_1(x) \rightleftharpoons \begin{cases} d(x) & \text{if } x \in X\setminus\{x_0\}, \\ \max\{d(f), d(g)\} + 1 & \text{if } x = x_0. \end{cases}$$

Then

$$(f_{x_0})_d = f_{d_1}. \tag{8}$$

Indeed, let N be the degree of f with respect to x_0, and let the coefficient at a monomial M in f be nonzero. If the degree of M with respect to x_0 equals N, then $d_1(M) \geq N \cdot d_1(x_0) > (N-1) \cdot d_1(x_0) + d(f)$. On the other hand, if the degree of M is less than N, then $d_1(M) \leq (N-1) \cdot d_1(x_0) + d(f)$. Therefore, (8) is true, and by Proposition 1.4 and Lemma 1.5 the following equality holds:

$$g = \sum_\mu u_\mu f v_\mu, \qquad d_1(u_\mu) + d_1(f) + d_1(v_\mu) \leq d_1(g)$$

(here "homogeneous" means "homogeneous with respect to d_1"). However, the inequality $m + d_1(f) + n \leq d_1(g)$ cannot be true for $m, n \geq 0$, since $d_1(g) = d(g) < d_1(x_0) \leq d_1(f)$. Therefore, $g = 0$.

Theorem 3 is proved.

The remainder of Chapter 1 is dedicated to the proof of Theorem 1.

§2. Reduction of Theorem 1 to a problem about subspaces of the field of rational functions

LEMMA 2.1. *A finite sublattice $\mathscr{L}$ in the lattice of subspaces of a linear space U over a field K is distributive if and only if it has a common basis.*

PROOF. If $\mathscr{L}$ has a common basis, then it is isomorphic to a sublattice in a lattice of subsets of this basis and therefore is distributive.

Conversely, let $\mathscr{L}$ be distributive. Use induction on $|\mathscr{L}|$. If $|\mathscr{L}|$ equals 1, the statement is obvious. Let $\mathscr{L}$ contain more than one element. Let V be the maximal element of $\mathscr{L}$ with respect to inclusion, and let V_0 be a maximal element among all the elements different from V. By the induction hypothesis, the lattice $\mathscr{L} \rightleftharpoons \{L \in \mathscr{L} \mid L \subseteq V_0\}$ has a common basis $\mathscr{B}_0$. Let

$$L_0 \rightleftharpoons \bigcap_{\substack{L \in \mathscr{L} \\ L + V_0 = V}} L = \bigcap_{\substack{L \in \mathscr{L} \\ L \not\subseteq V_0}} L.$$

Let $\mathscr{B}_1$ be a basis of the subspace L_0 modulo V_0. By the distributive law,

$$V_0 + L_0 = V_0 + \bigcap_{\substack{L \in \mathscr{L} \\ L + V_0 = V}} L = \bigcap_{\substack{L \in \mathscr{L} \\ L + V_0 = V}} (L + V_0) = V,$$

and therefore, the set $\mathscr{B} \rightleftharpoons \mathscr{B}_0 \cup \mathscr{B}_1$ is a basis of V. We shall prove that it is a common basis for $\mathscr{L}$. Indeed, let $\mathscr{B}'$ be a finite subset of $\mathscr{B}$, let L be an element of $\mathscr{L}$, and let $\sum_{b \in \mathscr{B}'} \lambda_b \cdot b \in L$, $\lambda_b \in K\setminus\{0\}$. If $\mathscr{B}' \cap \mathscr{B}_1 \neq \emptyset$, then $L \not\subseteq V_0$, and therefore, $L \supseteq L_0 \supseteq \mathscr{B}' \cap \mathscr{B}_1$. Hence, we can assume that $\mathscr{B}' \subseteq \mathscr{B}_0$. But then

$$\sum_{b \in \mathscr{B}'} \lambda_b \cdot b \in L \cap V_0 \in \mathscr{L}_0, \qquad \mathscr{B}' \subseteq L \cap V_0 \subseteq L.$$

The lemma is proved.

Lemma 2.1 implies the equivalence of the statements in Theorem 1 formulated at the beginning of this chapter.

Later we shall need the following

LEMMA 2.2. *Let U_1, U_2 be linear spaces over an arbitrary field K, and let $\mathscr{A}_1$, $\mathscr{A}_2$ be finite sets generating distributive sublattices in the lattices of subspaces of respective spaces U_1 and U_2. Then the set*

$$\mathscr{A} \rightleftharpoons \{A_1 \otimes_K A_2 \mid A_1 \in \mathscr{A}_1,\ A_2 \in \mathscr{A}_2\}$$

generates a distributive sublattice in the lattice of subspaces of the space $U \rightleftharpoons U_1 \otimes_K U_2$.

PROOF. If $\mathscr{B}_1$, $\mathscr{B}_2$ are common bases for sets $\mathscr{A}_1$, $\mathscr{A}_2$, respectively, then the set $\mathscr{B} \rightleftharpoons \{b_1 \otimes b_2 \mid b_1 \in \mathscr{B}_1,\ b_2 \in \mathscr{B}_2\}$ is obviously a common basis of the set $\mathscr{A}$.

The lemma is proved.

Let $U = U_1 \oplus U_2$ be a direct sum of abelian groups, and $\pi_i : U \to U_i$, $i = 1, 2$, be the natural projections. A subgroup $A \subseteq U$ is called *decomposable* if $A = A\pi_1 + A\pi_2$.

LEMMA 2.3. *Let $\mathscr{A}$ be a set of decomposable subgroups in a group $U = U_1 \oplus U_2$, and assume $\mathscr{A}$ generates a distributive lattice. Then the set $\mathscr{A}\pi_1 \rightleftharpoons \{A\pi_1 \mid A \in \mathscr{A}\}$ also generates a distributive lattice.*

PROOF. If A_1, A_2 are decomposable subgroups, then the same is true for the subgroups $A_1 + A_2$ and $A_1 \cap A_2$. Therefore, the map π_1 induces a lattice homomorphism from the lattice of decomposable subgroups onto the lattice of subgroups of U_1.

The lemma is proved.

Now we reduce Theorem 1' to a problem about subspaces in the ring of polynomials in commuting variables.

We must prove that the set $\{F^m f F^{k-m} \mid 0 \leq m \leq k\}$, where f is a d-homogeneous element and k a nonnegative integer, generates a distributive sublattice in the lattice of subspaces of the algebra $F = k\langle x_1, x_2, x_3, \ldots \rangle$. (The number of generators need not be countable; in fact, we can even assume that all free generators are included into the presentation of f.)

Consider a free algebra $k\langle x, y \rangle$ with unity. The elements

$$z_i \rightleftharpoons x^{3d(x_i)-1} y^{2i} xy, \qquad i = 1, 2, \ldots$$

obviously generate a subalgebra isomorphic to the algebra F. Let d_x be the degree function on the algebra $k\langle x, y \rangle$ with respect to the variable x

(i.e., $d_x(x) = 1$, $d_x(y) = 0$). Then the map $x_i \to z_i$ can be extended to an embedding of graded algebras, and we shall identify the image of F under this embedding with F. For any $h \in F$ we have $d(h) = 3d_x(h)$.

Let F_1 be a subalgebra of $k\langle x, y \rangle$ generated by unity and all monomials starting with x. Denote by F_1^m the subspace of all d_x-homogeneous elements of degree m in F_1. Given that the image of F is identified with F, we have $F^m \subseteq F_1^{3m}$. Denote by N^m the subspace generated by all monomials belonging to the set $F_1^{3m} \setminus F^m$. Then

$$F_1^{3m} = F^m \oplus N^m. \tag{1}$$

LEMMA 2.4. *Let f be a homogeneous element of nonzero degree in F. Then for any $m, n \geq 0$*

$$F_1^{3m} f N^n + N^m f F_1^{3n} \subseteq N^{m+d(f)+n}$$

and the subspace $F_1^{3m} f F_1^{3n}$ is decomposable in the space (1).

PROOF. Let $\mathscr{Z}$ be a semigroup generated by unity and the elements z_i, $i = 1, 2, \ldots$. For any monomials u, v, w it is true that

$$(v, uvw \in \mathscr{Z}, v \neq 1) \Rightarrow u, w \in \mathscr{Z}.$$

Therefore $F_1^{3m} f N^n$, $N^m f F_1^{3n} \subseteq N^{m+d(f)+n}$. Now the decomposability of the subspace $F_1^{3m} f F_1^{3n}$ follows from the equalities

$$F_1^{3m} f F_1^{3n} = F^m f F^n + (F_1^{3m} f N^n + N^m f F_1^{3n})$$
$$= F_1^{3m} f F_1^{3n} \cap F + F_1^{3m} f F_1^{3n} \cap N^{m+d(f)+n}.$$

The lemma is proved.

Therefore, it suffices to prove the distributivity of the lattice generated by a set $\{F_1^m f F_1^{k-m} \mid 0 \leq m \leq k\}$, where $f \in F_1$ is homogeneous.

Let $R \rightleftharpoons k[t_i \mid i \in \mathbb{Z}]$ be the ring of polynomials in the commuting variables t_i, where i runs over all integers. Let $\overline{R} \rightleftharpoons k(t_i \mid i \in \mathbb{Z})$ be the quotient field of R, let α be an automorphism of $\overline{R}$ mapping t_i to t_{i+1}, $\overline{R}[x, \alpha]$ be the ring of skew polynomials with the multiplication defined by the formula

$$t_i x = x t_i^\alpha = x t_{i+1}.$$

It is well known (see, for instance [9, section 0.7]) that elements x and t_0 generate a free subalgebra in the algebra $\overline{R}[x, \alpha]$ over the field k. We shall identify $k\langle x, y \rangle$ with its image under the monomorphism $x \to x$, $y \to t_0$.

It is clear that for $a \in \overline{R}$ the element $f = x^n a \in \overline{R}[x, \alpha]$ belongs to the subalgebra F_1 if and only if $a \in k[t_0, t_1, \ldots, t_{n-1}]$. Besides, the subspace $F_1^m f F_1^{k-n}$ in $\overline{R}[x, \alpha]$ has the form

$$x^{k+n} \cdot (k[t_0, \ldots, t_{k-m-1}] \cdot a^{\alpha^{k-m}} \cdot k[t_{k-m+n}, \ldots, t_{k-m-1}]).$$

Therefore, to prove Theorem 1 it is enough to show that the set of subspaces

$$\{k[t_0, \ldots, t_{i-1}] \cdot a^{\alpha^i} \cdot k[t_{n+i}, \ldots, t_{n+k-1}] \mid i = 0, 1, \ldots, k\}, \qquad (2)$$
$$a \in k[t_0, \ldots, t_{n-1}], \qquad n, k \text{ nonnegative integers,}$$

generates a distributive sublattice.

§3. Distributive sublattices of modular lattices

Let $\langle \mathscr{L}; +, \cap \rangle$ be an arbitrary modular lattice, let $\mathscr{A}$ be a subset of $\mathscr{L}$. A set $\mathscr{A}$ is called *distributive* if

$$A_0 \cap \sum_{i=1}^n A_i = \sum_{i=1}^n A_0 \cap A_i$$

for any $A_0, A_1, \ldots, A_n \in \mathscr{A}$. Denote by $\langle \mathscr{A} \rangle$ the set of all finite intersections of elements from $\mathscr{A}$.

LEMMA 3.1. *A set $\mathscr{A}$ generates a distributive lattice in $\mathscr{L}$ if and only if $\langle \mathscr{A} \rangle$ is distributive.*

PROOF. If $\mathscr{A}$ generates a distributive lattice, then $\langle \mathscr{A} \rangle$ is contained in this lattice and therefore is distributive.

Let $\langle \mathscr{A} \rangle$ be distributive. It suffices to prove that the set of all finite sums of elements from $\langle \mathscr{A} \rangle$ is distributive. This assertion is proved by induction in the number of summands using the following identities:

$$\begin{aligned}(A + B) \cap (C + D) &= (A + B) \cap ((A + B + C) \cap (C + D)) \\ &= (A + B) \cap (C + (A + B + C) \cap D) \\ &= (A + B) \cap (C + A \cap D + B \cap D) \\ &= A \cap D + B \cap D + (A + B) \cap C.\end{aligned}$$

The lemma is proved.

LEMMA 3.2. *A finite subset $\mathscr{A}$ of $\mathscr{L}$ is distributive if and only if any subset $\mathscr{A}' \subseteq \mathscr{A}$ contains an element A_0 such that*

$$A_0 \cap \sum_{A \in \mathscr{A}' \setminus \{A_0\}} A = \sum_{A \in \mathscr{A}' \setminus \{A_0\}} A_0 \cap A. \qquad (1)$$

PROOF. If $\mathscr{A}$ is distributive, then any element of $\mathscr{A}'$ satisfies (1). Conversely, suppose that the condition of the lemma holds. We use induction in the number $|\mathscr{A}|$. Obviously, we can assume that all proper subsets of $\mathscr{A}$ are distributive, and it is enough to prove that $A_1 \cap \sum_{A \in \mathscr{A} \setminus \{A_1\}} A = \sum_{A \in \mathscr{A} \setminus \{A_1\}} A_1 \cap A$ for any $A \in \mathscr{A}$. Let $A_2 \rightleftharpoons \sum_{A \in \mathscr{A} \setminus \{A_1, A_0\}} A$, where A_0 is the element satisfying (1) for $\mathscr{A}' = \mathscr{A}$. Then

$$A_1 \cap \sum_{A \in \mathscr{A} \setminus \{A_1\}} a = A_1 \cap (A_0 + A_2) = A_1 \cap ((A_1 + A_2) \cap (A_0 + A_2))$$

$$= A_1 \cap (A_2 + (A_1 + A_2) \cap A_0)$$
$$= A_1 \cap (A_2 + A_1 \cap A_0) = A_1 \cap A_2 + A_1 \cap A_0$$
$$= \sum_{A \in \mathscr{A} \setminus \{A_1\}} A_1 \cap A$$

by the distributivity of the subsets $\mathscr{A} \setminus \{A_0\}$, $\mathscr{A} \setminus \{A_1\}$, and the modularity of the lattice $\mathscr{L}$.

The lemma is proved.

Denote the set $\{A \cap B \mid A \in \mathscr{A}, B \in \mathscr{B}\}$, $\mathscr{A}, \mathscr{B} \subseteq \mathscr{L}$ by $\mathscr{A} \wedge \mathscr{B}$.

LEMMA 3.3. *Let $\mathscr{A}$ be a finite subset in $\mathscr{L}$, $\mathscr{U}, \mathscr{V} \subseteq \mathscr{A}$, $\mathscr{U} \neq \varnothing$, $W \in \mathscr{L}$, and $U \subseteq W$ for any $U \in \mathscr{U}$. Define*

$$\mathscr{A}^+ \rightleftharpoons (\mathscr{A} \setminus \mathscr{U}) \cup \{W\}, \quad \mathscr{A}^- \rightleftharpoons (\mathscr{A} \setminus \mathscr{V}) \cup (\mathscr{V} \wedge \{W\}).$$

If the sets $\mathscr{A}^+$ and $\mathscr{A}^-$ generate distributive lattices, then the set $\mathscr{A}$ also generates a distributive lattice.

PROOF. Let $\mathscr{A}' \subseteq \langle \mathscr{A} \rangle$. By Lemmas 3.1 and 3.2 it suffices to find an element $A_0 \in \mathscr{A}'$ satisfying (1). If W does not contain elements from the set $\mathscr{A}'$, then $\mathscr{A}' \subseteq \langle \mathscr{A} \setminus \mathscr{U} \rangle \subseteq \langle \mathscr{A}^+ \rangle$; hence, $\mathscr{A}'$ is distributive. Let $A_0 \in \mathscr{A}'$, $A_0 \subseteq W$. Then

$$A_0 \cap \sum_{A \in \mathscr{A}' \setminus \{A_0\}} A = A_0 \cap \left(W \cap \sum_{A \in \mathscr{A}' \setminus \{A_0\}} A \right)$$

$$= A_0 \cap \left(\sum_{\substack{A \in \mathscr{A}' \setminus \{A_0\} \\ A \subseteq W}} A + W \cap \sum_{\substack{A \in \mathscr{A}' \\ A \not\subseteq W}} A \right)$$

$$= A_0 \cap \sum_{A \in \mathscr{A}' \setminus \{A_0\}} (W \cap A). \qquad (2)$$

We can omit the parentheses in the last equality since $\{A \in \mathscr{A}' \mid A \subseteq W\} \subseteq \langle \mathscr{A}^+ \rangle$. Define

$$\mathscr{A}'_1 \rightleftharpoons \{A \in \mathscr{A}' \setminus \{A_0\} \mid (\forall V \in \mathscr{V}) \, A \not\subseteq V\},$$
$$\mathscr{A}'_2 \rightleftharpoons \{A \in \mathscr{A}' \setminus \{A_0\} \mid (\exists V \in \mathscr{V}) \, A \subseteq V\}.$$

The set $\{A_0\} \cup \mathscr{A}'_1 \cup (\mathscr{A}'_2 \wedge \{W\})$ is contained in $\langle \mathscr{A}^- \rangle$; hence, the equalities

(2) can be continued as follows:

$$= A_0 \cap \left(\sum_{A \in \mathscr{A}_1'} W \cap A + \sum_{A \in \mathscr{A}_2'} W \cap A \right)$$

$$= A_0 \cap \left(\sum_{A \in \mathscr{A}_1'} A + \sum_{A \in \mathscr{A}_2'} W \cap A \right) \cap \left(\sum_{A \in \mathscr{A}_1'} W \cap A + \sum_{A \in \mathscr{A}_2'} W \cap A \right)$$

$$= \left(\sum_{A \in \mathscr{A}_1'} A_0 \cap A + \sum_{A \in \mathscr{A}_2'} A_0 \cap A \right) \cap \left(\sum_{A \in \mathscr{A}_1'} W \cap A + \sum_{A \in \mathscr{A}_2'} W \cap A \right)$$

$$= \sum_{A \in \mathscr{A}' \setminus \{A_0\}} A_0 \cap A.$$

The lemma is proved.

A set $\mathscr{A} \subseteq \mathscr{L}$ is called an *antichain* if it does not contain a nontrivial chain.

LEMMA 3.4. *Let $\mathscr{A}$ be a finite subset in $\mathscr{L}$ and let each antichain in $\mathscr{A}$ generate a distributive lattice. Then $\mathscr{A}$ also generates a distributive lattice.*

PROOF. Let $A_1, A_2 \in \mathscr{A}$, $A_1 \subset A_2$. We can assume that the sets $\mathscr{A} \setminus \{A_1\}$ and $\mathscr{A} \setminus \{A_2\}$ generate distributive lattices. Define $\mathscr{U} \rightleftharpoons \{A_1\}$, $\mathscr{V} \rightleftharpoons \mathscr{A} \setminus \{A_1\}$, $W \rightleftharpoons A_2$ and apply Lemma 3.3. Clearly, $\mathscr{A}^+ = \mathscr{A} \setminus \{A_1\}$, and it suffices to prove the assertion for the set $\mathscr{A}^- = \mathscr{A} \setminus \{A_2\}$. We shall apply Lemmas 3.1 and 3.2. Let $\mathscr{A}' \subseteq \langle \mathscr{A}^- \rangle = \langle \mathscr{A} \rangle \wedge \{A_2\}$. If $A \nsubseteq A_1$ for any $A \in \mathscr{A}'$, then $\mathscr{A}'$ is contained in $\langle \mathscr{A} \setminus \{A_1\} \rangle$, and hence, is distributive. Let $A_0 \in \mathscr{A}'$, $A_0 \subseteq A_1$. Define

$$\mathscr{A}_1' \rightleftharpoons \{A \in \mathscr{A}' \mid A \subseteq A_1\} \setminus \{A_0\}, \qquad \mathscr{A}_2' \rightleftharpoons \{A \in \mathscr{A}' \mid A \nsubseteq A_1\}.$$

Then

$$A_0 \cap \sum_{A \in \mathscr{A}' \setminus \{A_0\}} A = A_0 \cap \left(A_1 \cap \left(\sum_{A \in \mathscr{A}_1'} A + \sum_{A \in \mathscr{A}_2'} A \right) \right) = A_0 \cap \left(\sum_{A \in \mathscr{A}_1'} A + A_1 \cap \sum_{A \in \mathscr{A}_2'} A \right).$$

If $A_2 \in \mathscr{A}_2'$, then $A_1 \subseteq \sum_{A \in \mathscr{A}_2'} A$ and $A_0 \cap \sum_{A \in \mathscr{A}' \setminus \{A_0\}} A = A_0$. Let $A_2 \notin \mathscr{A}_2'$. Then the set $\mathscr{A}_2'$ is of the form $\mathscr{A}_2'' \wedge \{A_2\}$, where $\mathscr{A}_2'' \subseteq \langle \mathscr{A} \setminus \{A_1, A_2\} \rangle$. By distributivity of the lattices generated by $\mathscr{A} \setminus \{A_1\}$ and $\mathscr{A} \setminus \{A_2\}$ we have

$$A_1 \cap \sum_{A \in \mathscr{A}_2''} A \cap A_2 = A_1 \cap \left(\sum_{A \in \mathscr{A}_2''} A \right) \cap A_2 = \sum_{A \in \mathscr{A}_2''} A_1 \cap A = \sum_{A \in \mathscr{A}_2'} A_1 \cap A,$$

$$A_0 \cap \sum_{A \in \mathscr{A}' \setminus \{A_0\}} A = A_0 \cap \sum_{A \in \mathscr{A}' \setminus \{A_0\}} A_1 \cap A = \sum_{A \in \mathscr{A}' \setminus \{A_0\}} A_0 \cap A.$$

The parentheses here can be opened since $\mathscr{A}' \wedge \{A_1\} \subseteq \langle \mathscr{A} \setminus \{A_2\} \rangle$.

The lemma is proved.

§4. Basic rings and basic modules

The plan for the completion of proof of Theorem 1 is as follows. We shall construct a "formal calculus", whose "formulas" are sets of subspaces of a special form. "Rules of inference" are certain special cases of Lemmas 2.2, 3.3, and 3.4, and one additional simple "rule". "Axioms" are sets of at most two elements: such sets always generate distributive lattices. Analyzing the "syntacsis" of a certain fragment of the constructed calculus, we shall bring to contradiction the assumption that for the set in question (given by (2) in §2) the inference does not exist.

This section contains the description of the "formal language" and of "rules of inference", and the proof of one "metatheorem" (Lemma 4.3). It is divided into three subsections.

1. In the first subsection, we give the main definitions and notations used in this and subsequent sections.

Let $\langle \mathscr{A} ; \subseteq \rangle$ be an arbitrary partially ordered set. Define the following preorder relation on the set of its subsets: $\mathscr{A}_1 \preceq \mathscr{A}_2$ if $(\forall A_1 \in \mathscr{A}_1) (\exists A_2 \in \mathscr{A}_2) A_1 \subseteq A_2$. A subset $\mathscr{A}_1 \subseteq \mathscr{A}$ is called an *antichain* if no two elements of $\mathscr{A}_1$ are comparable. Obviously, the relation $\preceq$ defines an order relation on the set of antichains in $\mathscr{A}$.

Denote the set of integers by $\mathbb{Z}$. Let $\lambda, \mu \in \mathbb{Z}$. Introduce the notations: $\langle \lambda, \mu \rangle \rightleftharpoons \{i \in \mathbb{Z} \mid \lambda \leq l < \mu\}$, $\langle \lambda, \infty \rangle \rightleftharpoons \{i \in \mathbb{Z} \mid \lambda \leq i\}$, $\langle -\infty, \mu \rangle \rightleftharpoons \{i \in \mathbb{Z} \mid i < \mu\}$, $\langle -\infty, \infty \rangle \rightleftharpoons \mathbb{Z}$. These sets are called *intervals*. Any nonempty interval I uniquely determines elements $\lambda(I), \mu(I) \in \mathbb{Z} \cup \{-\infty, \infty\}$ such that $I = \langle \lambda(I), \mu(I) \rangle$.

Let I_1, I_2 be two intervals not containing each other. Then either $\lambda(I_1) < \lambda(I_2)$, $\mu(I_1) < \mu(I_2)$ or $\lambda(I_1) > \lambda(I_2)$, $\mu(I_1) > \mu(I_2)$. In the first case we say that $I_1 < I_2$, in the second, $I_1 > I_2$. Therefore, on each antichain of intervals (with respect to inclusion) we define the order relation $<$. Let us introduce the following notation:

- $k[t_i \mid i \in \mathbb{Z}]$ is the ring of polynomials in commuting variables over the field k.
- $k(t_i \mid i \in \mathbb{Z})$ is the quotient field of $k[t_i \mid i \in \mathbb{Z}]$.
- $[X] \rightleftharpoons k[t_i \mid i \in X]$ is the subring of all polynomials in the variables t_i, $i \in X$, where X is a nonempty subset of $\mathbb{Z}$.
- (X) is the quotient field of the ring $[X]$.
  For $X = \varnothing$, we shall set $[X] = (X) = k$.
- $(J) \rightleftharpoons \prod_{I \in J}(I)$ is the product of subfields (I) of the field $(\mathbb{Z})$, where the product is taken over all the intervals in a set J.
- $[X]^*$ is the multiplicative semigroup of nonzero elements of $[X]$.
- $[J]^* \rightleftharpoons \prod_{I \in J}[I]^* = \{\prod_{I \in J} a_I \mid a_I \in [I]^*, I \in J\}$ is a product of semigroups $[I]^*$, where the product is taken over all intervals in a set J.

FREE ASSOCIATIVE ALGEBRAS AND INVERTING HOMOMORPHISMS OF RINGS   21

- $J(X)$ is a set of intervals contained in X ($X \subseteq \mathbb{Z}$).
- $s(a) \rightleftharpoons \bigcap_{I \in J(\mathbb{Z}), a \in [I]} I$ is the minimal interval containing the indices of all variables occurring in a polynomial $a \in [\mathbb{Z}]$. We call $s(a)$ the *support* of an element a. For $a \in k$ we define $s(a) = \varnothing$.
- $J(a) \rightleftharpoons \{s(q) \mid q \text{ is a prime divisor of } a\} \cup \{\varnothing\}$ is the collection of supports of all prime divisors of a polynomial a. (Recall that the ring $[\mathbb{Z}]$ is a unique factorization domain.)
- $(a)_J$ is the greatest divisor b of an element a such that $J(b) \subseteq J$ (J is a set of intervals).
- $a \wedge b$, $a \vee b$ are the greatest common divisor and the least common multiple of polynomials a and b.
- $\mathcal{T}_0^0 \rightleftharpoons \{[\{i\}] \mid i \in \mathbb{Z}\} \cup \{(I) \mid I \in J(\mathbb{Z})\}$.
- $\mathcal{T}_0 \rightleftharpoons \{\prod_{i=1}^m A_i \mid A_i \in \mathcal{T}_0^0, \, m = 1, 2, \ldots\}$ is a collection of finite products of all subfields of $(\mathbb{Z})$, belonging to $\mathcal{T}_0^0$.
- $\mathcal{T} \rightleftharpoons \{a \cdot A_0 \mid a \in [\mathbb{Z}], \, A_0 \in \mathcal{T}_0\}$.

Elements of the sets $\mathcal{T}_0$ and $\mathcal{T}$ are called *basic rings* and *basic modules*, respectively.

Let $A = a \cdot [X] \cdot (J)$ be a basic module, where $a \in [\mathbb{Z}]$, $J \subseteq J(\mathbb{Z})$, $|J|, |X| < \infty$. Denote

$$o(A) \rightleftharpoons [X] \cdot (J), \tag{1}$$

$$X(A) \rightleftharpoons X \cup \left(\bigcup_{I \in J} I\right), \tag{2}$$

$$J(A) \rightleftharpoons \bigcup_{I \in J} J(I), \tag{3}$$

$$h(A) \rightleftharpoons (a)_{J(a) \setminus J(A)}. \tag{4}$$

A few simple corollaries of the above definitions are:

1.1. $o(A) = [X] \cdot (J) = (J(A)) \cdot [X(A)] = [X(A)]([J(A)]^*)^{-1} = \{r/s \mid r \in [X(A)], s \in [J(A)]^*\}$.

1.2. $b^{-1} \in o(A) = [X] \cdot (J) \Leftrightarrow b \in [J(A)]^* \Leftrightarrow J(b) \subseteq J(A) \Leftrightarrow b \cdot o(A) = o(A) \Longrightarrow b \in o(A) \, (b \in [\mathbb{Z}])$.

1.3. $A = a \cdot o(A) = (a)_{J(a) \setminus J(A)} \cdot (a)_{J(A)} \cdot o(A) = h(A) \cdot o(A) = h(A) \cdot [X(A)]([J(A)]^*)^{-1} = \{h(A) \cdot r/s \mid r \in [X(A)], J(s) \subseteq J(A)\}$.

1.4. An irreducible fraction r/s belongs to A if and only if $h(A)$ divides r, $r/h(A) \in [X(A)]$, $J(s) \subseteq J(A)$.

1.5. For any basic module A the numbers $h(A)$, $o(A)$, $X(A)$, $J(A)$ do not depend on a representation of A in the form $a \cdot [X] \cdot (J)$. Indeed, 1.3 and 1.4 imply that

$$h(A) = \bigwedge_{r \in [\mathbb{Z}] \cap A} r \tag{5}$$

is a greatest common divisor of the polynomials in A,

$$o(A) = \{r/h(A) \mid r \in A\}, \tag{6}$$

$$X(A) = \{i \in \mathbb{Z} \mid t_i \in o(A)\}, \tag{7}$$

$$J(A) = \bigcup_{s \in [\mathbb{Z}],\, 1/s \in o(A)} J(s). \tag{8}$$

1.6. If $A_1, A_2 \in \mathscr{T}$, then $A_1 \cdot A_2 \in \mathscr{T}$, and also,

$$o(A_1 \cdot A_2) = o(A_1) \cdot o(A_2), \tag{9}$$

$$X(A_1 \cdot A_2) = X(A_1) \cup X(A_2), \tag{10}$$

$$J(A_1 \cdot A_2) = J(A_1) \cup J(A_2). \tag{11}$$

The statements follow from 1.3, (7), and (8).

1.7. If $A_1, A_2 \in \mathscr{T}_0$, i.e., $A_1, A_2 \in \mathscr{T}$, $h(A_1) = h(A_2) = 1$, then

$$X(A_1 \cap A_2) = X(A_1) \cap X(A_2), \tag{12}$$

$$J(A_1 \cap A_2) = J(A_1) \wedge J(A_2). \tag{13}$$

(as before, $\mathscr{A} \wedge \mathscr{B}$ stands for $\{A \cap B \mid A \in \mathscr{A}, B \in \mathscr{B}\}$ where $\mathscr{A}, \mathscr{B}$ are sets of subsets).

The statements follow from (6), (7), and (8).

1.8. The lattice $\langle \mathscr{T}_0; \cdot, \cap \rangle$ is distributive. This follows from (10)–(13).

1.9. Let $A_1, A_2 \in \mathscr{T}$. Then

1) if $A_1 \cap A_2 \neq \{0\}$, then

$$(h(A_1) \vee h(A_2))h(A_1)^{-1} \in [X(A_1)], \tag{14}$$

$$(h(A_1) \vee h(A_2))h(A_2)^{-1} \in [X(A_2)]. \tag{15}$$

2) If (14) and (15) hold, then $\{0\} \neq A_1 \cap A_2 \in \mathscr{T}$, and

$$h(A_1 \cap A_2) = h(A_1) \vee h(A_2), \tag{16}$$

$$o(A_1 \cap A_2) = o(A_1) \cap o(A_2). \tag{17}$$

PROOF. 1) Let r/s be an irreducible fraction contained in $A_1 \cap A_2$. By 1.4, we have $r = (h(A_1) \vee h(A_2))r_1$, $r_1 \in [\mathbb{Z}]$ and (14), (15) are true by (5).

2) Set $A \rightleftharpoons (h(A_1) \vee h(A_2)) \cdot (o(A_1) \cap o(A_2))$. Then by (4) and (13)

$$J(h(A_1) \vee h(A_2)) \cap J(A) = (J(h(A_1)) \cup J(h(A_2))) \cap (J(A_1) \wedge J(A_2))$$
$$\subseteq (J(h(A_1)) \cap J(A_1)) \cup (J(h(A_2)) \cap J(A_2)) = \{\varnothing\},$$

i.e., $h(A) = h(A_1) \vee h(A_2)$.

Now we apply 1.2. Let $a = h(A)$, $r/s \in A$, where $r \in [X(A)]$, $J(s) \subseteq J(A)$, $r \wedge s = 1$. Then $a \in A_1$ due to 1.4, (14), (13), and similarly, $a \in A_2$. Therefore, $A \subseteq A_1 \cap A_2$.

Let an irreducible fraction r/s belong to $A_1 \cap A_2$. By 1.4, $h(A_i) \mid r$, $i = 1, 2$, and consequently $h(A) \mid r$. Similarly, due to 1.4, (14), and (15), $r/h(A) \in [X(A)]$ and $J(s) \subseteq J(A)$; hence, $A_1 \cap A_2 = A$. The statement is proved.

1.10. Let $A_1, A_2 \in \mathcal{T}$. Then
$$A_1 \subseteq A_2 \Leftrightarrow (h(A_1)/h(A_2) \in [X(A_2)], \ o(A_1) \subseteq o(A_2)).$$

The statement follows immediately from 1.9.

1.11. Let $A_1, A_2 \in \mathcal{T}_0$. Then the homomorphism
$$\varepsilon(A_1, A_2): A_1 \otimes_{A_1 \cap A_2} A_2 \longrightarrow A_1 \cdot A_2$$
that maps $a_1 \otimes a_2$ to $a_1 \cdot a_2$ is an isomorphism.

PROOF. We use elementary properties of tensor products and quotient rings (see, for instance, [6, Chapter II, 2.7]). If A_i are polynomial rings, the statement is obvious.

Letting $X_i \rightleftharpoons X(A_i)$, $S_i \rightleftharpoons [J(A_i)]^*$, $i = 1, 2$, $X \rightleftharpoons X_1 \cup X_2$, we obtain the sequence of natural homomorphisms

$$\begin{aligned}
A_1 \cdot A_2 &= ([X]S_1^{-1})S_2^{-1} \xrightarrow{\sim} [X]S_1^{-1} \otimes_{[X_2]} [X_2]S_2^{-1} \\
&\xrightarrow{\sim} ([X] \otimes_{[X_1]} [X_1]S_1^{-1}) \otimes_{[X_2]} [X_2]S_2^{-1} \\
&\xrightarrow{\sim} [X_1]S_1^{-1} \otimes_{[X_1]} [X] \otimes_{[X_2]} [X_2]S_2^{-1} \\
&\xrightarrow{\sim} [X_1]S_1^{-1} \otimes_{[X_1]} ([X_1] \otimes_{[X_1 \cap X_2]} [X_2]) \otimes_{[X_2]} [X_2]S_2^{-1} \\
&\xrightarrow{\sim} ([X_1]S_1^{-1} \otimes_{[X_1]} [X_1]) \otimes_{[X_1 \cap X_2]} ([X_2] \otimes_{[X_2]} [X_2]S_2^{-1}) \\
&\xrightarrow{\sim} [X_1]S_1^{-1} \otimes_{[X_1 \cap X_2]} [X_2]S_2^{-1} \xrightarrow{\sim} [X_1]S_1^{-1} \otimes_{[X_1 \cap X_2](S_1 \cap S_2)^{-1}} [X_2]S_2^{-1} \\
&= A_1 \otimes_{A_1 \cap A_2} A_2.
\end{aligned}$$

The last equality follows immediately from 1.7 and 1.3. It is also rather easy to see that the composition of these isomorphisms is the inverse map to ε. The statement is proved.

Let $A \in \mathcal{T}$ be a basic module, and let $I^1, I^2 \in \mathcal{T}(\mathbb{Z})$ be two intervals. Define the basic module
$$\pi(A, I^1, I^2) \rightleftharpoons (A \cdot (I^2)) \cap (I^1). \tag{18}$$

Let us assume that
$$(I^1 \cap I^2) \subseteq o(A), \qquad J(h(A)) \preceq \{I^1, I^2\}. \tag{19}$$

Then $J(h(A)) \subseteq J(I^1) \cup J(I^2)$, $J(h(A)) \cap J(I^1 \cap I^2) = \varnothing$ (since the supports of prime divisors $h(A)$ do not belong to the set $J(A) \supseteq J(I^1 \cap I^2)$). It follows immediately that

$$h(\pi(A, I^1, I^2)) = h(A))_{J(I^1)}, \qquad h(\pi(A, I^2, I^1)) = (h(A))_{J(I^2)}, \tag{20}$$
$$o(\pi(A, I^1, I^2)) = o(A) \cap (I^1), \qquad o(\pi(A, I^2, I^1)) = o(A) \cap (I^2). \tag{21}$$

Furthermore,
$$\begin{aligned}
\pi(A, I^1, I^2) \cdot \pi(A, I^2, I^1) &= (h(A))_{J(I^1)} \cdot (h(A))_{J(I^2)} \cdot (o(A) \cap (I^1)) \cdot (o(A) \cap (I^2)) \\
&= h(A) \cdot (o(A) \cap (I^1)(I^2)) = A \cap (I^1)(I^2).
\end{aligned}$$

The basic modules $\pi(A, I^1, I^2)$ and $\pi(A, I^2, I^1)$ are subspaces of the respective spaces (I^1) and (I^2) over the field $(I^1 \cap I^2)$. Applying 1.11, we get the following result.

1.12. Let $A \in \mathcal{T}$, $I^1, I^2 \in J(\mathbb{Z})$, and let the conditions
$$(I^1 \cap I^2) \subseteq o(A) \subseteq (I^1)(I^2), \qquad J(h(A)) \preceq \{I^1, I^2\}$$
be satisfied. Then
$$\varepsilon((I^1), (I^2))\big(\pi(A, I^1, I^2) \otimes_{(I^1 \cap I^2)} \pi(A, I^2, I^1)\big) = A.$$

1.13. Let $\mathcal{A} \subseteq \mathcal{T}$, $|\mathcal{A}| < \infty$, $I^1, I^2 \in J(\mathbb{Z})$, and let the conditions
$$(I^1 \cap I^2) \subseteq o(A) \subseteq (I^1)(I^2), \qquad J(h(A)) \preceq \{I^1, I^2\}$$
hold for any $A \in \mathcal{A}$. If the sets
$$\pi(\mathcal{A}, I^1, I^2) \rightleftharpoons \{\pi(A, I^1, I^2) \mid A \in \mathcal{A}\},$$
$$\pi(\mathcal{A}, I^2, I^1) \rightleftharpoons \{\pi(A, I^2, I^1) \mid A \in \mathcal{A}\}$$
generate distributive lattices, then the set $\mathcal{A}$ also generates a distributive lattice. This statement immediately follows from 1.12 and Lemma 2.2.

1.14. Let $\mathcal{A} \in \mathcal{T}$. Then

1) The set $c^{-1}\mathcal{A} \rightleftharpoons \{c^{-1}A \mid A \in \mathcal{A}\}$ is contained in $\mathcal{T}$ if and only if for any $(\forall A \in \mathcal{A})$ we have $(c)_{J(c) \setminus J(A)} \mid h(A)$.

2) $c^{-1}\mathcal{A}$ generates a distributive lattice if and only if $\mathcal{A}$ generates a distributive lattice.

PROOF. Multiplication by c is an automorphism of the space $(\mathbb{Z})$, hence the second assertion follows. The first one follows from
$$c^{-1}A = ((c)_{J(c) \setminus J(A)})^{-1} h(A) \cdot o(A)$$
for any $A \in \mathcal{T}$.

2. In this subsection we give the definition of inference, that will serve as means for obtaining distributive lattices from other distributive lattices.

A finite partially ordered set $\mathcal{D}$ is called a *tree* if any two of its elements possess a lower bound, and no two elements possess an upper bound. Elements of $\mathcal{D}$ are called *nodes*. The minimal element is called the *initial node*, maximal elements are called *terminal nodes*.

Let $d \in \mathcal{D}$. Define the germ at the node d to be a subset $\{d, d_1, \ldots, d_n\}$, where $d_1, \ldots, d_n$ are all the nodes immediately following d. The notation for the germ will be:

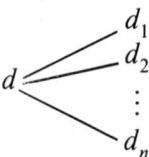

An *inference* is a tree with nodes corresponding to finite sets of nonzero basic modules such that each germ of this tree has one of the four forms listed below:

I. Let $\mathscr{A} \subseteq \mathscr{T} \setminus \{\{o\}\}$, $W \in \mathscr{T}$, $\mathscr{U}, \mathscr{V} \subseteq \mathscr{A}$, and let the conditions

$$\mathscr{U} \neq \varnothing, \quad \mathscr{U} \preceq \{W\} \tag{I.1}$$

hold.
Define

$$F^+(\mathscr{A}, \mathscr{U}, W) \rightleftharpoons (\mathscr{A} \setminus \mathscr{U}) \cup \{W\},$$
$$F^-(\mathscr{A}, \mathscr{V}, W) \rightleftharpoons (\mathscr{A} \setminus \mathscr{V}) \cup (\mathscr{V} \wedge \{W\}) \setminus \{\{o\}\}.$$

Then the germ

$$\mathscr{A} \diagdown \begin{matrix} F^+(\mathscr{A}, \mathscr{U}, W) \\ F^-(\mathscr{A}, \mathscr{V}, W) \end{matrix}$$

is called a type I germ or, alternatively, inference rule I.

II. Let $\mathscr{A} \subseteq \mathscr{T}$, $I^1, I^2 \in J(\mathbb{Z})$, and let the conditions

$$(I^1 \cap I^2) \subseteq o(A) \subseteq (I^1)(I^2), \tag{II.1}$$
$$J(h(A)) \preceq \{I^1, I^2\} \tag{II.2}$$

hold for any $A \in \mathscr{A}$.
Then the germ

$$\mathscr{A} \diagdown \begin{matrix} \pi(\mathscr{A}, I^1, I^2) \\ \pi(\mathscr{A}, I^2, I^1) \end{matrix}$$

is called a type II germ.

III. Let $\mathscr{A} \subseteq \mathscr{T}$, $c \in [\mathbb{Z}]$, and let the condition

$$(c)_{J(c) \setminus J(A)} \mid h(A) \tag{III.1}$$

hold for any $A \in \mathscr{A}$. The germ

$$\mathscr{A} \longrightarrow c^{-1} \mathscr{A}$$

is called a type III germ.

IV. Let $\mathscr{A} \subseteq \mathscr{T}$, $A_1, A_2 \in \mathscr{A}$, $A_1 \subset A_2$. The germ

$$\mathscr{A} \diagdown \begin{matrix} \mathscr{A} \setminus \{A_1\} \\ \mathscr{A} \setminus \{A_2\} \end{matrix}$$

is called a type IV germ.

A finite set $\mathscr{A}$ is said to be *provable* if there exists an inference such that its initial node is $\mathscr{A}$ and all its terminal nodes are sets of at most two elements.

By Lemmas 2.2, 3.3, 3.4, and statements 1.13, 1.14 each provable set of basic modules generates a distributive lattice (under the operations $\langle +, \cap \rangle$).

Let us establish a few simple properties of the objects we have defined.

2.1. *Let $\mathscr{D}$ be an inference, let I be an interval, and let the initial node of $\mathscr{D}$ satisfy the condition*

$$(\forall A)\ h(A) \in [I]. \tag{22}$$

Then all the nodes of $\mathscr{D}$ satisfy this condition.

PROOF. It suffices to show that the condition holds for one inference rule. For the inference rule I it is a consequence of 1.9 and of $h(W)|h(A)$ for some $A \in \mathscr{A}$; for rule II it follows from (20), for rule III from $h(c^{-1}A)\ |\ h(A)$.

2.2. *Let $\mathscr{A}, \mathscr{U}, \mathscr{V}, W$ define a germ of type I, and let I be an interval satisfying* (22). *Let*

$$\mathscr{A}_1 \rightleftharpoons \mathscr{A} \wedge \{(I)\},\quad \mathscr{U}_1 \rightleftharpoons \mathscr{U} \wedge \{(I)\},\quad \mathscr{V}_1 \rightleftharpoons \mathscr{V} \wedge \{(I)\},\quad W_1 \rightleftharpoons W \cap (I).$$

Then the sets $\mathscr{A}_1, \mathscr{U}_1, \mathscr{V}_1, W$ define a germ of type I, and

$$F^+(\mathscr{A}_1, \mathscr{U}_1, W_1) \subseteq F^+(\mathscr{A}, \mathscr{U}, W) \wedge \{(I)\},$$
$$F^-(\mathscr{A}_1, \mathscr{V}_1, W_1) \subseteq F^-(\mathscr{A}, \mathscr{V}, W) \wedge \{(I)\}.$$

The statements follow directly from definitions.

2.3. *Let $\mathscr{A}, I^1, I^2$ define a germ of type II, and let I be an interval satisfying* (22). *Let*

$$\mathscr{A}_1 \rightleftharpoons \mathscr{A} \wedge \{(I)\},\qquad I_1^1 \rightleftharpoons I^1 \cap I,\qquad I_1^2 \rightleftharpoons I^2 \cap I.$$

Then the sets $\mathscr{A}_1, I_1^1, I_1^2$ define a germ of type II and

$$\pi(\mathscr{A}_1, I_1^1, I_1^2) \subseteq \pi(\mathscr{A}, I^1, I^2) \wedge \{(I)\}. \tag{23}$$

PROOF. Indeed, (II.1) and (II.2) are obvious, and the inclusion (23) is a direct consequence of (20), (21).

2.4. *Let $\mathscr{A}, c$ define a germ of type III, and let I be an interval satisfying* (22). *Then $\mathscr{A} \wedge \{(I)\}, (c)_{J(I)}$ define a germ of type III, and*

$$((c)_{J(I)})^{-1} \cdot (\mathscr{A} \wedge \{(I)\}) \subseteq (c^{-1}\mathscr{A}) \wedge \{(I)\}.$$

PROOF. By (22) and (III.1) we have $(\forall A \in \mathscr{A}) J(c)\backslash J(A) \preceq \{I\}$; hence,

$$(c)_{J(c)\backslash J(A)} = ((c)_{J(I)})_{J(c)\wedge\{I\})\backslash J(A\cap(I))}$$

for any $A \in \mathscr{A}$. Therefore, (III.1) holds for the pair $(\mathscr{A} \wedge \{(I)\}, (c)_{J(I)})$.

2.5. *A subset of a provable set is also provable.*

PROOF. Let $\mathscr{D}$ be an inference with the initial node $\mathscr{A}$, and let $\mathscr{A}_1$ be a subset of the set $\mathscr{A}$. By induction on the inference length it is easy to construct an inference $\mathscr{D}_1$ with the initial node $\mathscr{A}_1$ such that its nodes are subsets of the nodes of the inference $\mathscr{D}$.

2.6. *Let I be an interval, and let $\mathscr{D}$ be an inference such that its initial node $\mathscr{A}$ satisfies (22). Then the set $\mathscr{A} \wedge \{(I)\}$ is provable.*

PROOF. By induction on the inference length, using 2.1–2.5.

3. In this subsection we prove the following result.

LEMMA 4.3. *Let $\mathscr{A}$ be a finite set of basic modules, $J = \{I_1 < \cdots < I_m\}$ be a finite antichain of nonempty intervals. Define*

$$T_0(\mathscr{A}, J) \rightleftharpoons \{A \in \mathscr{A} \mid o(A) \supseteq (J) = \prod_{i=1}^{m}(I_i)\},$$

$$T_{I_i}(\mathscr{A}, J) \rightleftharpoons \{A \in \mathscr{A} \mid o(A) \not\supseteq (I_i)\}, \quad i = 1, \ldots, m.$$

We assume that the following conditions are satisfied:
1) $T_0(\mathscr{A}, J) \subseteq \mathscr{T}_0$;
2) $(\forall A \in \mathscr{A})(A \in T_{I_i}(\mathscr{A}, J) \Rightarrow h(A) \in [I_i])$;
3) $T_{I_i}(\mathscr{A}, J) \cap T_{I_j}(\mathscr{A}, J) = \varnothing$, $i, j \in \{1, \ldots, m\}$, $i \neq j$;
4) *The sets* $T_{I_i}(\mathscr{A}, J) \wedge \{(I_i)\}$ *are provable for all* $i \in \{1, \ldots, m\}$. *Then the set $\mathscr{A}$ is provable.*

PROOF. Note that if $\mathscr{A} \subseteq \mathscr{T}_0$, the conditions of the lemma are satisfied for any antichain J of the form $J = \{I_1\}$, $I_1 = \{i\}$, $i \in \mathbb{Z}$. Indeed, in this case $\mathscr{A} \wedge \{(I_1)\} \subseteq \mathscr{T}_0 \wedge \{(I_1)\} = \{(\{i\}), [\{i\}], [\varnothing]\}$.

Let $A \in \mathscr{T}_0^0$. Define

$$\Gamma(A) \rightleftharpoons \begin{cases} \{i\} & \text{if } A = [\{i\}], \\ \{\lambda(I), \mu(I) - 1\} & \text{if } A = (I), \ I \neq \varnothing, \\ \varnothing & \text{if } a = [\varnothing]. \end{cases}$$

Let

$$A = \prod_{i=1}^{k} A_i \in \mathscr{T}_0, \qquad A_i \in \mathscr{T}_0^0. \tag{24}$$

Since each A_i is a ring with unity, we can assume the set $\{A_i \mid i = 1, \ldots, k\}$ to be an antichain; in this case the representation (24) is called *irreducible*. Uniqueness of the irreducible representation follows from 1.8.

Let $A = \prod_{i=1}^{k} A_i$ be the irreducible representation of a basic ring A as a product of rings from $\mathscr{T}_0^0$. Define $\Gamma(A) \rightleftharpoons \bigcup_{i=1}^{k} \Gamma(A_i)$. For $A \in \mathscr{T}$, define $\Gamma(A) \rightleftharpoons \Gamma(o(A))$; for $\mathscr{A} \subseteq \mathscr{T}$ define $\Gamma(\mathscr{A}) \rightleftharpoons \bigcup_{A \in \mathscr{A}} \Gamma(A)$. Let J be a finite nonempty anticain of nonempty intervals. Define

$$\Gamma(\mathscr{A}, J) \rightleftharpoons \Gamma(\mathscr{A}) \cup \Gamma((J)), \qquad \gamma(\mathscr{A}, J) \rightleftharpoons |\Gamma(\mathscr{A}, J)|,$$
$$\tau(\mathscr{A}, J) \rightleftharpoons |(\mathscr{A} \setminus T_0(\mathscr{A}, J)|.$$

Let the conditions of the lemma be satisfied, and let $\gamma(\mathscr{A}, J) = 1$. Then $J = \{I_1\}$, $I_1 = \{i\}$, $\mathscr{A} \subseteq \{(I_1)\} \cup T_{I_1}(\mathscr{A}, J) \wedge \{(I_1)\}$, and $\mathscr{A}$ is provable by 4), 2.5, and rule IV. Hence, we can assume that the lemma is true for any pair $(\mathscr{A}', J)$ satisfying both the conditions of the lemma and the inequality $\gamma(\mathscr{A}', J') < \gamma(\mathscr{A}, J)$.

We are going to prove, by induction on τ with γ fixed, the existence of an inference $\mathscr{D}(\mathscr{A}, J)$ with the initial node $\mathscr{A}$ and every terminal node either being an axiom (i.e., having at most two elements) or satisfying the conditions

$$\Gamma(\mathscr{A}') \subseteq \Gamma(\mathscr{A}, J), \qquad \mathscr{A}' = T_0(\mathscr{A}', J). \qquad (25)$$

From now on $T_i(\mathscr{A}, J) \rightleftharpoons T_{i_i}(\mathscr{A}, J)$.

If either of the two conditions

$$\mathscr{A} = T_0(\mathscr{A}, J) \qquad (\exists i \in \{1, \ldots, m\})\mathscr{A} = T_i(\mathscr{A}, J) \wedge \{(I_i)\} \qquad (26)$$

holds, the proof is trivial, since in the first case the inference $\mathscr{D}(\mathscr{A}, J)$ consists of a single node $\mathscr{A}$, and in the second case the set $\mathscr{A}$ is provable by 4).

We shall prove later that if neither of the conditions in (26) holds, then there exist intervals I^1, I^2 satisfying

$$\Gamma((I^1)(I^2)) \subseteq \Gamma(\mathscr{A}, J), \qquad \Gamma(\mathscr{A}, J) \not\subseteq I^i, \qquad i = 1, 2, \qquad (27)$$

$$J \preceq \{I^1, I^2\}, \qquad (28)$$

$$(\forall A \in \mathscr{A}) \, o(A) \supseteq (I^1 \cap I^2), \qquad (29)$$

$$(\exists A \in \mathscr{A} \setminus T_0(\mathscr{A}, J)) \, A \subseteq (I^1)(I^2). \qquad (30)$$

Given two intervals I^1, I^2 satisfying (27)–(30), define $W \rightleftharpoons (I^1)(I^2)$, $\mathscr{U} \rightleftharpoons \{A \in \mathscr{A} \mid A \subseteq W\}$, $\mathscr{V} \rightleftharpoons \mathscr{A} \setminus \mathscr{U}$, and apply rule I. The set $\mathscr{A}^+ \rightleftharpoons F^+(\mathscr{A}, \mathscr{U}, W)$ and the antichain J satisfy the conditions of the lemma:

$$T_0(\mathscr{A}^+, J) = \{W\} \cup (T_0(\mathscr{A}, J) \cap \mathscr{V}), \qquad T_i(\mathscr{A}^+, J) = T_i(\mathscr{A}, J) \cap \mathscr{V},$$
$$i = 1, \ldots, m;$$

the first three conditions are obvious, the fourth follows from 2.5. By (27) and (30) $\Gamma(\mathscr{A}^+, J) \subseteq \Gamma(\mathscr{A}, J)$, $\tau(\mathscr{A}^+, J) < \tau(\mathscr{A}, J)$, and the inference $\mathscr{D}(\mathscr{A}^+, J)$ exists by the induction assumption. Hence, it suffices to show that the set $\mathscr{A}^- \rightleftharpoons F^-(\mathscr{A}, \mathscr{V}, W) = \mathscr{A} \wedge \{W\}$ is provable. By 2), (28), (29), this set satisfies the conditions (II.1) and (II.2) required for applying rule II.

Let us verify that the conditions of the lemma hold for the sets $\pi(\mathscr{A}^-, I^1, I^2)$ and $\pi(\mathscr{A}^-, I^2, I^1)$. By (20), $\pi(\mathscr{A} \wedge \{W\}, I^1, I^2) = \pi(\mathscr{A}, I^1, I^2) \rightleftharpoons \mathscr{A}^1$.

If $A \in T_0(\mathscr{A}, J)$, then $\pi(A, I^1, I^2) = A \cap (I^1)$ by (21). If $A \in T_i(\mathscr{A}, J)$, $i \in \{1, \ldots, m\}$, $I_i \not\subseteq I^1$, then 2), (28), (29), and (20) yield

$$J(I^1) \wedge J(I_i) \subseteq J(I^1) \wedge J(I^2) = J(I^1 \cap I^2) \subseteq J(o(A)) = J(A),$$
$$h(\pi(A, I^1, I^2)) = (h(A))_{J(I^1)} = (h(A))_{J(I^1) \wedge J(I_i)} = (h(A))_{J(A)} = 1.$$

Using (21) and the last equality, we obtain $\pi(A, I^1, I^2) = o(A) \cap (I^1)$. If $A \in T_i(\mathscr{A}, J)$, $I_i \subseteq I^1$, then $\pi(A, I^1, I^2) = h(A) \cdot (o(A) \cap (I^1))$.

Therefore, if $(\forall i \in \{1, \ldots, m\})$ $I_i \not\subseteq I^1$, then $\pi(\mathscr{A}, I^1, I^2) \subseteq \mathscr{T}_0$, and as we noted in the beginning of the proof, the conditions of the lemma hold for the pair $\mathscr{A}^1 = \pi(\mathscr{A}, I^1, I^2)$, J^1, where $J^1 = \{I\}$, $I = \{\lambda\}$, $\lambda \in \Gamma(\mathscr{A}, J) \cap I^1 \cap \mathbb{Z}$ (if the last set is empty, then $I^1 = \varnothing$, $\mathscr{A} = \{[\varnothing]\}$). On the other hand, if $I_i \subseteq I^1$ for some $i \in \{1, \ldots, m\}$, then let $J^1 \rightleftharpoons \{I_i \mid I_i \subseteq I^1\}$. We have

$$T_0(\mathscr{A}^1, J^1) = \left\{ \pi(A, I^1, I^2) \mid A \in T_0(\mathscr{A}, J) \cup \left(\bigcup_{I_i \not\subseteq I^1} T_i(\mathscr{A}, J) \right) \right\},$$

$$T_i(\mathscr{A}^1, J^1) = \pi(T_i(\mathscr{A}, J), I^1, I^2), \qquad I_i \subseteq I^1.$$

The conditions of the lemma are obviously carried over from $T_i(\mathscr{A}, J)$ to $T_i(\mathscr{A}^1, J^1)$. In both cases we have $\Gamma(\mathscr{A}^1, J^1) \subseteq \Gamma(\mathscr{A}, J) \cap I^1$, and the set $\mathscr{A}^1$ is provable by (27) and the induction assumption with respect to γ. Similarly, the set $\mathscr{A}^2 \rightleftharpoons \pi(\mathscr{A}, I^2, I^1)$ is provable, and the desired inference $\mathscr{D}(\mathscr{A}, J)$ is constructed.

Next we shall prove that if neither of the conditions (26) holds, then the intervals I^1, I^2 satisfying (27)–(30) indeed exist.

Since the first equality in (26) is false, there exists $k \in \{1, \ldots, m\}$ such that $T_k \neq \varnothing$. We have $\Gamma(\mathscr{A}, J) \not\subseteq I_k$, since otherwise the second part of (26) would be true. Therefore, either $\alpha_0 \geq \mu(I_k)$, or $\alpha_0 < \lambda(I_k)$ for some $\alpha_0 \in \Gamma(\mathscr{A}, J)$. We consider only the first case, since the intervals I^1, I^2 are constructed in a completely similar way in both cases. Let

$$I^1 \rightleftharpoons \langle \min \Gamma(\mathscr{A}, J), \mu(I_k) \rangle,$$
$$\mu_2 \rightleftharpoons \max \Gamma(\mathscr{A}, J) + 1,$$
$$\lambda_2(A) \rightleftharpoons \max\{\lambda \in \mathbb{Z} \cup \{+\infty\} \mid (I^1)(\langle \lambda, \mu_2 \rangle) \supseteq A\}, \qquad (A \in T_k),$$
$$\lambda_2 \rightleftharpoons \max\{\lambda_2(A) \mid A \in T_k\},$$
$$I^2 \rightleftharpoons \langle \lambda_2, \mu_2 \rangle.$$

Verify (27)–(30). Let A_0 be an element of T_k such that $\lambda_2 = \lambda_2(A_0)$. Then $A_0 \subseteq (I^1)(I^2)$ which implies (30). Furthermore, $(I^1)(I^2) \supseteq (I_k)$, $(I^1)(I^2) \supseteq o(A_0) \supseteq \prod_{j \neq k}(I_j)$ by 3). This implies (28). Next observe that $\Gamma((I^1)(I^2)) \subseteq \Gamma(\mathscr{A}, J)$, $\alpha_0 \notin I^1$. If $I^1 \cap I^2 = \varnothing$, then $\lambda(I_k) \in \Gamma(\mathscr{A}, J) \setminus I^2$, i.e., (27) holds, and (29) is obvious. So we shall assume $I^1 \cap I^2 \neq \varnothing$. By definition of λ_2, $o(A_0) \not\subseteq (I^1)(\langle \lambda_2 + 1, \mu_2 \rangle) \supseteq [\Gamma(\mathscr{A}, J)]$. Therefore,

$$(I) \subseteq o(A_0), \qquad \{I\} \not\leq \{I^1, \langle \lambda_2 + 1, \mu_2 \rangle\}, \qquad I \subseteq I^1 \cup \langle \lambda_2 + 1, \mu_2 \rangle$$

for some interval I. Hence, $I \supseteq \langle \lambda_2, \mu(I_k) + 1 \rangle$, and if $\lambda_2 \leq \lambda(I_k)$ then

$(I_k) \subseteq (I) \subseteq o(A_0)$, $A_0 \notin T_k$, which is impossible. Consequently, $\lambda_2 > \lambda(I_k)$, $I^1 \cap I^2 \subseteq I_k$, $\lambda(I_k) \notin I^2$, and (27) is verified.

Let us verify (29). If $A \in \mathscr{A} \setminus T_k$, then $o(A) \supseteq (I_k) \supseteq (I^1 \cap I^2)$. On the other hand, if $A \in T_k$, then $o(A) \not\subseteq (I^1)(\langle \lambda_2(A)+1, \mu_2 \rangle) \supseteq (I^1)(\langle \lambda_2+1, \mu_2 \rangle)$; hence, $o(A) \supseteq (\langle \lambda_2, \mu(I_k)+1 \rangle) \supseteq (I^1 \cap I^2)$.

Next we extend the resulting inference $\mathscr{D}(\mathscr{A}, J)$ to an inference of the set $\mathscr{A}$. Given a set $\mathscr{A}'$ satisfying (25), we want to prove that $\mathscr{A}'$ is provable. Let I_1' be the maximal interval satisfying

$$\lambda(I_1'), \mu(I_1') - 1 \in \Gamma(\mathscr{A}, J), \qquad \Gamma(\mathscr{A}, J) \not\subseteq I_1'.$$

Such an interval always exists with the exception of the case $\Gamma(\mathscr{A}, J) \subseteq \{-\infty, \infty\}$, but in this case $\mathscr{A} \subseteq \{[\varnothing], (\mathbb{Z})\}$ and there is nothing to prove.

Let $J' \rightleftharpoons \{I_1'\}$. Then the pair $\mathscr{A}', J'$ satisfies the conditions of the lemma (the fourth condition is satisfied by the induction assumption about the value of γ). Consider a terminal node $\mathscr{A}''$ of the inference $\mathscr{D}(\mathscr{A}', J')$ consisting of more than two elements. The set $\mathscr{A}''$ satisfies (25); moreover, $|\Gamma(\mathscr{A}'', J') \setminus I_1'| \leq 1$ and it is easy to see that $\mathscr{A}''$ is an antichain. Therefore, it is provable.

The lemma is proved.

COROLLARY. *Every finite set of basic rings is provable.*

§5. Corollaries of Lemma 4.3

1. From now on, the symbol Ω with subscripts denotes finite subsets of $\mathbb{Z}$, and $\mathbb{Z}_-^\nu$, $\mathbb{Z}_+^\nu$ denote the respective intervals $\langle -\infty, \nu \rangle$, $\langle \nu, \infty \rangle$. Let $\lambda, \mu \in \mathbb{Z}$, $\lambda < \mu$, $a \in [\langle \lambda, \mu \rangle]$. Define $I_0 \rightleftharpoons \langle \lambda, \mu \rangle$, $N_0 \rightleftharpoons (\mathbb{Z}_-^\lambda)(\mathbb{Z}_+^\mu)$. Throughout this section we shall assume λ, μ, a to be fixed.

As was shown in §2, to prove Theorem 1 it suffices to prove that for any set Ω and any finite interval J such that $\Omega + I_0 \subseteq J$ the set

$$\{(aN_0)^{\alpha^\omega} \cap [J] \mid \omega \in \Omega\} \tag{1}$$

generates a distributive lattice (here we are referring to statements 1.9 and 1.7 from §4).

LEMMA 5.1. *Assume that every set of the form*

$$\{(aN_0)^{\alpha^\omega} \mid \omega \in \Omega\} \tag{2}$$

is provable. For some Ω, let there be given a family of basic rings $\{A_\omega \mid \omega \in \Omega\}$ satisfying

$$(\forall \omega \in \Omega) N_0^{\alpha^\omega} \cap [\Omega + I_0] \subseteq A_\omega \subseteq N_0^{\alpha^\omega}. \tag{3}$$

Then the set

$$\mathscr{A} \rightleftharpoons \{a^{\alpha^\omega} A_\omega \mid \omega \in \Omega\} \tag{4}$$

generates a distributive lattice.

PROOF. The proof is split into several parts.

1. Clearly we can assume that the statement is proved for any proper subset $\Omega' \subset \Omega$ and any family $\{A'_\omega \mid \omega \in \Omega'\}$ satisfying the condition of the lemma.

2. If $\Omega + I_0$ is an interval, then the set $\{(aN_0)^{\alpha^\omega} \cap (\Omega + I_0) \mid \omega \in \Omega\}$ is provable by 4.2.6.

3. Let $J \rightleftharpoons \{I_1 < \cdots < I_m\}$ be a maximal (with respect to the partial ordering $\preceq$) antichain of intervals, contained in the set $\Omega + I_0$. Then $I_i \cap I_{i+1} = \varnothing$, $i = 1, \ldots, m-1$, $T_0(\mathscr{A}, J) = \varnothing$. Define $\Omega_i \rightleftharpoons \{\omega \in \Omega \mid \omega + I_0 \subseteq I_i\}$. Then $I_i = \Omega_i + I_0$, $i = 1, \ldots, m$.

Define the basic ring

$$N(\Omega, \omega) \rightleftharpoons N_0^{\alpha^\omega} \cap \prod_{i=1}^m (I_i).$$

4. If for every $\omega \in \Omega$ the condition

$$N(\Omega, \omega) \subseteq A_\omega \subseteq N_0^{\alpha^\omega} \tag{5}$$

is satisfied, then the set $\mathscr{A}$ is provable.

Let us verify the conditions of Lemma 4.3 for the pair $(\mathscr{A}, J)$. We have

$$T_0(\mathscr{A}, J) = \varnothing, \qquad T_i(\mathscr{A}, J) = \{a^{\alpha^\omega} A_\omega \mid \omega \in \Omega_i\}, \qquad i = 1, \ldots, m.$$

1) is trivial, 2) and 3) follow from part 3 of the proof, and 4) follows from parts 2) and 3).

5. Next, let $\{A_\omega\}$ be an arbitrary family satisfying (3). We use induction, at each step increasing one of the rings $A_\omega \cap N(\Omega, \omega)$ and leaving others unchanged. Since the sets Ω, I_0 are finite, after a finite number of steps, (5) will become true for any $\omega \in \Omega$.

Let $A_{\omega_1} \not\supseteq N(\Omega, \omega_1)$ for some $\omega_1 \in \Omega$. Then there exists a nonempty interval $I = \langle \lambda_1, \mu_1 \rangle$ such that

$$(I) \subseteq N(\Omega, \omega_1), \qquad (I) \not\subseteq A_{\omega_1}, \qquad (\langle \lambda_1 + 1, \mu_1 \rangle) \subseteq A_{\omega_1}$$

(possibly with $\langle \lambda_1 + 1, \mu \rangle = \varnothing$). Condition 1) implies $I \subseteq \Omega + I_0$; hence, $\lambda_1 \in \omega_2 + I_0$ for some $\omega_2 \in \Omega$. Therefore, $t_{\lambda_1} \notin N_0^{\alpha^{\omega_2}}$, so that

$$N_0^{\alpha^{\omega_2}} \cap (\langle \lambda_1, \mu_1 \rangle) = N_0^{\alpha^{\omega_2}} \cap (\langle \lambda_1 + 1, \mu_1 \rangle) \subseteq A_{\omega_1}.$$

Furthermore,

$$(A_{\omega_1} \cdot (I)) \cap A_{\omega_2} = (A_{\omega_1} \cap A_{\omega_2}) \cdot ((I) \cap A_{\omega_2})$$
$$= (A_{\omega_1} \cap A_{\omega_2}) \cdot (N_0^{\alpha_2^\omega} \cap (I) \cap A_{\omega_2}) = A_{\omega_1} \cap A_{\omega_2},$$

and by 4.1.9,

$$a^{\alpha^{\omega_1}} A_{\omega_1}(I) \cap a^{\alpha^{\omega_2}} A_{\omega_2} = a^{\alpha^{\omega_1}} A_{\omega_1} \cap a^{\alpha^{\omega_2}} A_{\omega_2}. \tag{6}$$

Now we define

$$W \rightleftharpoons a^{\alpha^{\omega_1}} A_{\omega_1}(I), \quad U \rightleftharpoons a^{\alpha^{\omega_1}} A_{\omega_1}, \quad V \rightleftharpoons a^{\alpha^{\omega_2}} A_{\omega_2}, \quad \mathscr{U} \rightleftharpoons \{U\}, \quad \mathscr{V} \rightleftharpoons \{V\},$$

and apply Lemma 3.3. By the induction assumption, $\mathscr{A}^+ \rightleftharpoons F^+(\mathscr{A}, \mathscr{U}, W)$
$= (\mathscr{A}\setminus\{U\}) \cup \{W\}$ generates a distributive lattice.

By (6), the set $\mathscr{A}^- \rightleftharpoons F^-(\mathscr{A}, \mathscr{V}, W)$ contains the chain $\{U, V \cap W\}$. Consequently, each one of its antichains is contained in one of the following sets:

$$\mathscr{A}\setminus\{V\}, \quad (\mathscr{A}\setminus\{U, V\}) \cup \{V \cap W\}.$$

Both sets generate distributive lattices: the first due to part 1, and the second since it is contained in $\langle \mathscr{A}^+ \rangle$. Hence, the distributivity of the lattice generated by $\mathscr{A}$ follows from Lemma 3.4.

Lemma 5.1 is proved.

Thus, to prove Theorem 1 it remains to prove that sets of the form (2) are provable. In this section we shall prove that sets of a particular form are provable. These sets will be playing the role of "axioms" for inferring sets of the form (2).

2. We need several definitions and notations. Define the following relations on the set of subsets of a partially ordered set $\langle J(\mathbb{Z}); \subseteq \rangle$:

$$J_1 \vDash J_2 \rightleftharpoons (\forall I_1 \in J_1)(\forall I_2 \in J_2)\, I_1 \not\subset I_2,$$
$$J_1 \vdash J_2 \rightleftharpoons (\forall I_1 \in J_1)(\forall I_2 \in J_2)\, I_1 \not\subseteq I_2,$$
$$J_1 \dashv\vdash J_2 \rightleftharpoons (J_1 \vdash J_2, J_2 \vdash J_1).$$

Note that if J_1 and J_2 are antichains and $J_1 \dashv\vdash J_2$, then $J_1 \cap J_2 = \varnothing$ and $J_1 \cup J_2$ is also an antichain.

Let $I = \langle \rho, \sigma \rangle \in J(\mathbb{Z})$, $\rho, \sigma \in \mathbb{Z}$, $\rho < \sigma$. Denote by U_I the largest basic ring not containing the field (I). Evidently, such a ring exists, since if $A_1, A_2 \in \mathscr{T}_0$, $A_i \not\supseteq (I)$, $i = 1, 2$, then $A_1 \cdot A_2 \not\supseteq (I)$ by 4.1.6. It is rather easy to see that U_I has the form

$$U_I = (\mathbb{Z}_-^{\sigma-1}) \cdot [I](\mathbb{Z}_+^{\rho+1}). \tag{7}$$

If $\mathscr{A}$ and $\mathscr{B}$ are two sets, then $\mathscr{A} + \mathscr{B}$ always denotes $\{A + B \mid A \in \mathscr{A}, B \in \mathscr{B}\}$, where A and B can also be sets.

We introduce notation for sets of basic modules:

$$A^\Omega \rightleftharpoons \{A^{\alpha^\omega} \mid \omega \in \Omega\}, \quad A \in \mathscr{T};$$
$$\mathscr{A}(\Omega, \{a_\omega\}) \rightleftharpoons \{a_\omega N_0^{\alpha^\omega} \mid \omega \in \Omega\}, \quad a_\omega \in [\omega + I_0], \quad \omega \in \Omega;$$
$$\mathscr{B}(\mathscr{M}, \{q_M\}) \rightleftharpoons \{q_M U_M \mid M \in \mathscr{M}\}.$$

In the last definition $\mathscr{M}$ is a finite antichain of nonempty intervals, the q_M are nonfactorizable elements of the ring $[M]$ satisfying the condition $s(q_M) = M$, and the basic rings U_M are defined by (7).

We shall consider sets of the form
$$\mathscr{C} \rightleftharpoons \mathscr{C}(\Omega, \{a_\omega\}, \mathscr{M}, \{q_M\}) \rightleftharpoons \mathscr{A}(\Omega, \{a_\omega\}) \cup \mathscr{B}(\mathscr{M}, \{q_M\}) \qquad (8)$$
and introduce special notation for the following conditions that will play an important role in what follows:

$$\Phi_1(a, \Omega, \{a_\omega\}) \rightleftharpoons (\forall \omega \in \Omega)\, a_\omega \mid a^{\alpha^\omega},$$
$$\Phi_2(\Omega, \{a_\omega\}, \mathscr{M}, \{q_M\}) \rightleftharpoons (\forall \omega \in \Omega)(\forall M \in \mathscr{M})\, q_M \nmid a_\omega,$$
$$\Phi_3(\Omega, \{a_\omega\}, \mathscr{M}) \rightleftharpoons (\forall \omega \in \Omega)\, \mathscr{M} \vDash J(a_\omega),$$
$$\Phi_4(\Omega, \{a_\omega\}) \rightleftharpoons (\forall \omega_1, \omega_2 \in \Omega)\, a_{\omega_1} N_0^{\alpha^{\omega_1}} \cap a_{\omega_2} N_0^{\alpha^{\omega_2}} \neq \{0\},$$
$$\Phi_5(\Omega, \mathscr{M}) \rightleftharpoons (\forall M \in \mathscr{M})\, |\Omega \cap (M - I_0)| \leq 1,$$
$$\Phi_6(\Omega, \mathscr{M}) \rightleftharpoons (\forall \omega \in \Omega)(\exists \nu \in \mathbb{Z})$$
$$\{\mathbb{Z}_-^\nu, \mathbb{Z}_+^\nu\} \succcurlyeq (\omega + \{\mathbb{Z}_-^\lambda, \mathbb{Z}_+^\mu\}) \cup \mathscr{M}.$$

We shall sometimes write $\Phi_i(\mathscr{C})$ for a set $\mathscr{C}$ of the form (8), or omit arguments of Φ_i if they are clear from the context.

LEMMA 5.2. *Every set of the form* (8) *satisfying* $\Phi_1 - \Phi_6$ *is provable.*

The proof breaks into several parts.

1. *If a set* $\mathscr{A}(\Omega, \{a_\omega\})$ *satisfies* Φ_4, *then*
$$\mathscr{A}(\Omega, \{a_\omega\}) = c_\Omega \cdot N_0^\Omega, \qquad c_\Omega = \bigvee_{\omega \in \Omega} a_\omega$$
(recall that by definition $c_\Omega \cdot N_0^\Omega = \{c_\omega N_0^{\alpha^\omega} \mid \omega \in \Omega\}$).
PROOF. Since $a_\omega \mid c_\Omega$, we have
$$h(a_\omega N_0^{\alpha^\omega}) = a_\omega \mid (c_\Omega)_{J(c_\Omega) \setminus J(N_0^{\alpha^\omega})} = h(c_\Omega N_0^{\alpha^\omega})$$
for every $\omega \in \Omega$. Therefore, it suffices to show that $h(c_\Omega N_0^{\alpha^\omega}) \mid a_\omega$. Indeed, otherwise there would exist an irreducible polynomial q and a nonnegative integer k such that $q^k \mid c_\Omega$, $q^k \nmid a_\omega$, $q \notin N_0^{\alpha^\omega}$. By definition of c_Ω, there exists a number $\omega_1 \in \Omega$ such that $q^k \mid a_{\omega_1}$. In this case $q \mid (a_\omega \vee a_{\omega_1}) \cdot a_\omega^{-1}$; hence, $(a_\omega \vee a_{\omega_1}) \cdot a_\omega^{-1} \notin N_0^{\alpha^\omega}$ and $a_\omega N_0^{\alpha^\omega} \cap a_{\omega_1} N_0^{\alpha^{\omega_1}} = \{0\}$ by 4.1.9. This contradicts Φ_4. The assertion is proved.

Therefore, when Φ_4 holds, Φ_3 can be rewritten as
$$\Phi_3'(\Omega, \{a_\omega\}, \mathscr{M}) \rightleftharpoons \mathscr{M} \vDash J(c_\Omega).$$
The notation c_Ω will be fixed until the end of the proof of Lemma 5.2.

2. *A set* $\mathscr{A}(\Omega, \{a_\omega\})$ *satisfying* Φ_4 *is provable.*

PROOF. Since the pair $(\mathscr{A}(\Omega, \{a_\omega\}), c_\Omega)$ satisfies (III.1), the set
$$c_\Omega^{-1} \mathscr{A}(\Omega, \{a_\omega\}) = N_0^\Omega \subseteq \mathscr{T}_0$$
is provable by the Corollary to Lemma 4.3.

3. *A set of the form* (8) *satisfying* Φ_1-Φ_6 *and*

$$\Phi_5'(\Omega, \mathcal{M}) \rightleftharpoons (\forall M \in \mathcal{M})\ \Omega \cap (M - I_0) = \varnothing$$

is provable.

PROOF. Let $J \rightleftharpoons \{I_1 \prec \cdots \prec I_m\}$ be a maximal (with respect to the order $\preccurlyeq$) antichain of intervals that are contained in $\Omega + I_0$. Denote also $\Omega_i \rightleftharpoons \{\omega \in \Omega \mid \omega + I_0 \subseteq I_i\}$, $i = 1, \ldots, m$. By Φ_5', $J \cup \mathcal{M}$ is an antichain; we shall verify that it satisfies the conditions of Lemma 4.3. We have

$$T_0(\mathcal{C}, J \cup \mathcal{M}) = \varnothing;$$
$$T_{I_i}(\mathcal{C}, J \cup \mathcal{M}) = \mathcal{A}(\Omega_i, \{a_\omega\}), \qquad i = 1, \ldots, m;$$
$$T_M(\mathcal{C}, J \cup \mathcal{M}) = \{q_M U_M\}, \qquad M \in \mathcal{M}.$$

The conditions 1)–3) are evident. For T_{I_i}, $i = 1, \ldots, m$, the condition 4) follows from part 2 and statement 4.2.6, and for T_M, $M \in \mathcal{M}$, it is satisfied since T_M is a one-element set.

4. Next, consider the general case. Let $\mathcal{C}$ be a set of the form (8) satisfying Φ_1-Φ_6. We shall argue by induction, at each step decreasing one of the two sets $\Omega, \mathcal{M}$ without increasing the other one. After a finite number of steps Φ_5' will become true.

Assume that Φ_5' is not true. Choose a number $\omega_0 \in \Omega$ such that

$$M \cap (\omega_0 + I_0) \neq \varnothing \tag{9}$$

for some $M \in \mathcal{M}$. From the sets of the form $\mathcal{N}_{\rho,\sigma} \rightleftharpoons \{\mathbb{Z}_-^\rho, \mathbb{Z}_+^\sigma\}$, $\rho < \sigma$ satisfying

$$\mathcal{N}_{\rho,\sigma} \succcurlyeq (\omega_0 + \{\mathbb{Z}_-^\lambda, \mathbb{Z}_+^\mu\}) \cup \mathcal{M} \tag{10}$$

select a minimal $\mathcal{N}_{\rho_0,\sigma_0}$ with respect to the order $\preccurlyeq$. By Φ_6, there always exist sets $\mathcal{N}_{\rho,\sigma}$ satisfying (10), and by (9) we have $\mathcal{N}_{\rho_0,\sigma_0} \not\succcurlyeq \omega_0 + \{\mathbb{Z}_-^\lambda, \mathbb{Z}_+^\mu\}$, i.e., one of the two inequalities: $R_0 > \omega_0 + \lambda$, $\sigma_0 < \omega_0 + \mu$, is true. Assume that the first one is true (in case of the second one, proceed symmetrically). Then $\{\mathbb{Z}_-^{\rho_0-1}, \mathbb{Z}_+^{\sigma_0}\} \succcurlyeq \omega_0 + \{\mathbb{Z}_-^\lambda, \mathbb{Z}_+^\mu\}$ and by the minimality of the set $\mathcal{N}_{\rho_0,\sigma_0}$, there exists an interval $M_0 \in \mathcal{M}$ such that $\mu(M_0) = \rho_0$. Let

$$W \rightleftharpoons a_{\omega_0}(\mathbb{Z}_-^{\rho_0-1})(\mathbb{Z}_+^{\rho_0}) = c_\Omega(\mathbb{Z}_-^{\rho_0-1})(\mathbb{Z}_+^{\rho_0}),$$
$$U \rightleftharpoons a_{\omega_0} N_0^{\alpha^{\omega_0}} = c_\Omega N_0^{\alpha^{\omega_0}}, \qquad \mathcal{U} \rightleftharpoons \{U\},$$
$$V \rightleftharpoons q_{M_0} U_{M_0}, \qquad \mathcal{V} \rightleftharpoons \{V\}.$$

Since $s(q_{M_0}) = M_0$ and $\mu(M_0) = \rho_0$, the polynomial q_{M_0} has a nonzero degree with respect to the variable t_{ρ_0-1}. Therefore, $q_{M_0} \notin [\mathbb{Z}_-^{\rho_0-1} \cup \mathbb{Z}_+^{\rho_0}] = [\mathbb{Z}\setminus\{\rho_0 - 1\}]$. Hence, $V \cap W = \{0\}$ by 4.1.9 and Φ_2. The set $\mathcal{C}^- \rightleftharpoons$

$F^-(\mathscr{C}, \mathscr{V}, W) = \mathscr{A}(\Omega, \{a_\omega\}) \cup \mathscr{B}(\mathscr{M}\backslash\{M_0\}, \{q_M\})$ is provable by the induction hypothesis. (Subsets of sets of the form (8) are themselves of the same form, and Φ_1-Φ_6 can be carried over to subsets.)

Let us establish that the set

$$\mathscr{C}^+ \rightleftharpoons F^+(\mathscr{C}, \mathscr{U}, W) = \mathscr{A}(\Omega\backslash\{\omega_0\}, \{a_\omega\}) \cup \mathscr{B}(\mathscr{M}, \{q_M\}) \cup \{c_\Omega(\mathbb{Z}_-^{p_0-1})(\mathbb{Z}_+^{p_0})\} \tag{11}$$

is provable. Let

$$J_1 \rightleftharpoons J(c_\Omega)\backslash(J(\mathbb{Z}_-^{p_0-1} \cup \mathbb{Z}_+^{p_0}) \cup \{M_0\}), \qquad c \rightleftharpoons (c_\Omega)_{J_1}. \tag{12}$$

Verify (III.1) for the pair $(\mathscr{C}^+, c)$, i.e., verify that

$$(\forall A \in \mathscr{C}^+)(c)_{J(c)\backslash J(A)} \mid h(A).$$

Consider several cases:

1) $A = a_\omega N_0^{\alpha^\omega}$, $\omega \in \Omega\backslash\{\omega_0\}$. In this case

$$(c)_{J(c)\backslash J(A)} = (c_\Omega)_{J_1 \backslash J(N_0^{\alpha^\omega})} \mid (c_\Omega)_{J(c_\Omega)\backslash J(N_0^{\alpha^\omega})} = a_\omega.$$

2) $A = W = c_\Omega(\mathbb{Z}_-^{p_0-1})(\mathbb{Z}_+^{p_0}) = (c_\Omega)_{J(c_\Omega)\backslash J(\mathbb{Z}_-^{p_0-1} \cup \mathbb{Z}_+^{p_0})} \cdot (\mathbb{Z}_-^{p_0-1})(\mathbb{Z}_+^{p_0})$. In this case

$$(c)_{J(c)\backslash J(\mathbb{Z}_-^{p_0-1}\cup \mathbb{Z}_+^{p_0})} \mid (c_\Omega)_{J(c_\Omega)\backslash J(\mathbb{Z}_-^{p_0-1}\cup\mathbb{Z}_+^{p_0})} = h(W),$$
$$J(h(c^{-1}W)) \subseteq \{M_0, \varnothing\}. \tag{13}$$

3) $A = q_M U_M$, $M \in \mathscr{M}$. If $M \neq M_0$, then by (10)

$$M \in J(\mathbb{Z}_-^{p_0-1} \cup \mathbb{Z}_+^{\sigma_0}) \subseteq J(\mathbb{Z}_-^{p_0-1} \cup \mathbb{Z}_+^{p_0}) \tag{14}$$

(we are using here that $\mathscr{M}$ is an antichain and that $\mu(M_0) = \rho_0$). Therefore, $M \notin J_1$. If on the other hand, $M = M_0$, then $M \notin J_1$ by (12). Hence, $\mathscr{M} \cap J_1 = \varnothing$. Moreover, $\mathscr{M} \vDash J_1$ by Φ_3'; hence, $\mathscr{M} \vdash J_1$, i.e., $I \not\supseteq M$ for any $I \in J_1$, $M \in \mathscr{M}$. Applying (7), we obtain $(J_1) \subseteq U_M$ for any $M \in \mathscr{M}$ and

$$(c)_{J(c)\backslash J(U_M)} = (c)_{\{\varnothing\}} = 1, \qquad M \in \mathscr{M},$$
$$c^{-1}\mathscr{B}(\mathscr{M}; \{q_M\}) = \mathscr{B}(\mathscr{M}, \{q_M\}).$$

Condition (III.1) is verified. It remains to establish that the set $\mathscr{C}' \rightleftharpoons c^{-1}\mathscr{C}^+$ is provable. Define $I_1 \rightleftharpoons \mathbb{Z}_-^{p_0-1}$, $I_2 \rightleftharpoons M_0$, $I_3 \rightleftharpoons \mathbb{Z}_+^{p_0}$, $J \rightleftharpoons \{I_1 < I_2 < I_3\}$ and verify the conditions of Lemma 4.3 for the pair $(\mathscr{C}, J)$. Clearly, $T_0(\mathscr{C}', J) = \varnothing$. Let

$$\Omega_i \rightleftharpoons \{\omega \in \Omega\backslash\{\omega_0\} \mid \omega + I_0 \subseteq I_i\}, \qquad i = 1, 3;$$
$$T_i \rightleftharpoons T_{I_i}(\mathscr{C}', J), \qquad i = 1, 2, 3.$$

Condition 1) of Lemma 4.3 is obvious. By Φ_5, $(\omega + I_0) \cap M_0 = \varnothing$ for any $\omega \in \Omega \setminus \{\omega_0\}$. Therefore,

$$(\Omega \setminus \{\omega_0\}) + \{I_0\} \preceq \{I_1, I_3\}, \qquad \Omega_1 \cup \Omega_3 = \Omega \setminus \{\omega_0\},$$

and since $\mathscr{M}$ is an antichain, from (10) we obtain

$$\mathscr{M} \setminus \{M_0\} \preceq \{I_1, I_3\}.$$

Hence,

$$T_i = \{c^{-1} a_\omega N_0^{\alpha^\omega} \mid \omega \in \Omega_i\} \cup \{q_M U_M \mid M \in \mathscr{M}, M \subseteq I_i\}, \qquad i = 1, 3;$$
$$T_2 = \{q_{M_0} U_{M_0}, c^{-1} W\}$$

which proves condition 3). Condition 2) is obvious for T_1 and T_3 and follows from (13) for T_2. To verify 4), note that the set

$$T_1 \cup T_3 = \{c^{-1} a_\omega N_0^{\alpha^\omega} \mid \omega \in \Omega \setminus \{\omega_0\}\} \cup \mathscr{B}(\mathscr{M} \setminus \{M_0\}\{q_M\})$$

has the form (8), satisfies $\Phi_1 - \Phi_6$, and is provable by the induction assumption. The sets $T_i \wedge \{(I_i)\}$, $i = 1, 3$, are provable by 4.2.5 and 4.2.6. The set T_2 has two elements and so is provable.

The proof of Lemma 5.2 is completed.

§6. Special sets and special inference

All the notation and definitions of the previous section are carried over to the present section.

Consider a set

$$\mathscr{C} = \mathscr{C}(\Omega, \{a_\omega\}, \mathscr{M}, \{q_M\}) = \mathscr{A}(\Omega, \{a_\omega\}) \cup \mathscr{B}(\mathscr{M}, \{q_M\}). \qquad (1)$$

We say this set is *a-special* if it satisfies Φ_1, Φ_2, Φ_3. As noted in §5, it is rather easy to see that a subset of an *a*-special set is also *a*-special.

Let $\mathscr{C}$ be an *a*-special set of the form (1), q be an irreducible polynomial from $[\mathbb{Z}]$ satisfying

$$\Phi_7(\Omega, \{a_\omega\}, q) \rightleftharpoons (\exists \omega_1, \omega_2 \in \Omega)(q \mid a_{\omega_1}, q \nmid a_{\omega_2}, q \notin N_0^{\alpha^{\omega_2}});$$
$$\Phi_8(\omega, \{a_\omega\}, q) \rightleftharpoons (\forall \omega \in \Omega) \{s(q)\} \vDash J(a_\omega);$$
$$\Phi_9(\mathscr{M}, q) \rightleftharpoons \{s(q)\} \vdash \mathscr{M}.$$

Define

$$W \rightleftharpoons W(q) \rightleftharpoons q U_{s(q)},$$
$$\mathscr{U} \rightleftharpoons \mathscr{U}(\Omega, \{a_\omega\}, q) \rightleftharpoons \{a_\omega N_0^{\alpha^\omega} \mid \omega \in \Omega, q \mid a_\omega\}, \qquad (2)$$
$$\mathscr{V} \rightleftharpoons \mathscr{V}(\Omega, \{a_\omega\}, q) \rightleftharpoons \{a_\omega N_0^{\alpha^\omega} \mid \omega \in \Omega, (q \vee a_\omega) a_\omega^{-1} \notin N_0^{\alpha^\omega}\}. \qquad (3)$$

It is rather easy to see that

$$\mathscr{U} = \{A \in \mathscr{C} \mid A \subseteq W\}, \qquad \mathscr{V} = \{A \in \mathscr{C} \mid A \cap W = \{0\}\}.$$

The sets $\mathscr{U}$ and $\mathscr{V}$ are nonempty by Φ_7, and rule I is applicable. Let
$$F^+(\mathscr{C}, q) \rightleftharpoons F^+(\mathscr{C}, \mathscr{U}, W), \qquad F^-(\mathscr{C}, q) \rightleftharpoons F^-(\mathscr{C}, \mathscr{V}, W).$$
We call the germ
$$\mathscr{C} \begin{array}{c} F^+(\mathscr{C}, q) \\ F^-(\mathscr{C}, q) \end{array}$$
a *special germ of type* I or a *special rule* I.

The set $F^-(\mathscr{C}, q) = \mathscr{C} \setminus \mathscr{V}$ is obviously a-special. We shall prove that the set $\mathscr{C}^+ \rightleftharpoons F^+(\mathscr{C}, q)$ is a-special. We have
$$\begin{aligned} \mathscr{C}^+ &= \mathscr{A}(\Omega^+, \{a_\omega\}) \cup \mathscr{B}(\mathscr{M}, \{q_M\}) \cup \{qU_{s(q)}\}, \\ \Omega^+ &\rightleftharpoons \Omega^+(\mathscr{C}, q) \rightleftharpoons \{\omega \in \Omega \mid q \nmid a_\omega\}. \end{aligned} \qquad (4)$$
By Φ_3, Φ_7, there exists $\omega_1 \in \Omega$ such that
$$\mathscr{M} \vDash J(a_\omega) \ni s(q).$$
Therefore, $\mathscr{M} \vDash \{s(q)\}$, and Φ_9 implies that the set $\mathscr{M}^+ \rightleftharpoons \mathscr{M} \cup \{s(q)\}$ is an antichain. Setting $q_{s(q)} \rightleftharpoons q$ we obtain
$$\mathscr{C}^+ = \mathscr{A}(\Omega^+, \{a_\omega\}) \cup \mathscr{B}(\mathscr{M}^+, \{q_M\}).$$
Condition $\Phi_1(a, \Omega^+, \{a_\omega\})$ is obvious, $\Phi_2(\Omega^+, \{a_\omega\}, \mathscr{M}^+, \{q_M\})$ follows from $\Phi_2(\mathscr{C})$ and from the definition of Ω^+, and $\Phi_3(\Omega^+, \{a_\omega\}, \mathscr{M}^+)$ follows from $\Phi_3(\mathscr{C})$ and Φ_8.

Let $\mathscr{C}$ be an a-special set of the form (10) satisfying Φ_5 and let c be a polynomial satisfying
$$\begin{aligned} \Phi_{10}(\Omega, \{a_\omega\}, c) &\rightleftharpoons (\forall \omega \in \Omega)\, h(cN_0^{\alpha^\omega}) \mid a_\omega, \\ \Phi_{11}(\Omega, c) &\rightleftharpoons J(c) \preceq \{(\omega_1 + I_0) \cap (\omega_2 + I_0) \mid \omega_1, \omega_2 \in \Omega,\, \omega_1 \neq \omega_2\}. \end{aligned}$$
Let
$$c_\omega \rightleftharpoons h(cN_0^{\alpha^\omega}) = (c)_{J(c) \setminus J(N_0^{\alpha^\omega})}, \qquad \omega \in \Omega.$$
Now Φ_{10} yields $c_\omega^{-1} a_\omega \in [\omega + I_0]$ for any $\omega \in \Omega$; hence,
$$c^{-1} \mathscr{A}(\Omega, \{a_\omega\}) = \mathscr{A}(\Omega, \{c_\omega^{-1} a_\omega\}). \qquad (5)$$
Φ_5 and Φ_{11} immediately yield $\mathscr{M} \vdash J(c)$, which means
$$c^{-1} \mathscr{B}(\mathscr{M}, \{q_M\}) = \mathscr{B}(\mathscr{M}, \{q_M\}). \qquad (6)$$
Therefore, $c^{-1}\mathscr{C}$ is an a-special set, and we call the germ
$$\mathscr{C} \text{———} c^{-1}\mathscr{C}$$
a *special germ of type* III.

We say that an inference is *special* if all its germs are special (of type I or III). It has already been proved that if the initial node of a special inference is a-special, then so are all its terminal nodes.

LEMMA 6.1. *If the initial node of a special inference has the form* $\mathscr{A}(\Omega_1, \{a_\omega\})$ *then all its terminal nodes of the form* $\mathscr{C}(\Omega, \{a'_\omega\}, \mathscr{M}, \{q_M\})$ *satisfy the condition*

$$\Phi_{12}(\Omega_1, \Omega, \mathscr{M}) \rightleftharpoons (\forall M \in \mathscr{M})(\exists \omega_1, \omega_2 \in \Omega_1)$$
$$(\omega_1 \notin \Omega, \omega_1 \neq \omega_2, M \subseteq \omega_1 + I_0, M \cap (\omega_2 + I_0) \neq \varnothing).$$

PROOF. Since Φ_{12} is imposed only on sets Ω and $\mathscr{M}$, Φ_{12} continues to hold after applying special rule III. In case special rule I is applied, the set $F^-(\mathscr{C}, q)$ is contained in $\mathscr{C}$, and hence, satisfies Φ_{12} if $\mathscr{C}$ does; $F^+(\mathscr{C}, q)$ has the form (4), and as ω_1, ω_2 for $M = s(q)$ we take the elements that exist by $\Phi_7(\Omega, \{a_\omega\}, q)$.
The lemma is proved.

Let p be an irreducible polynomial belonging to the ring $[\mathbb{Z}]$. Introduce the notation:

$$\overline{p} \rightleftharpoons \{p^{\alpha^k} \mid k \in \mathbb{Z}\}, \qquad a_p \rightleftharpoons \prod_{q \in \overline{p}, \, q \mid a} q. \tag{7}$$

The next section contains the proof of the

MAIN LEMMA. *Let p be an irreducible divisor of an element a with the maximal value of $|s(p)|$, and let Ω_0 be a finite set of integers. Then there exists a special inference with nodes of the form*

$$\mathscr{A}(\Omega_1) \cup \mathscr{B}(\Xi), \tag{8}$$

where Ω_1, Ξ *are finite sets of integers,* $\Omega_1 \subseteq \Omega_0$,

$$\mathscr{A}(\Omega_1) \rightleftharpoons (aN_0)^{\Omega_1}, \qquad \mathscr{B}(\Xi) \rightleftharpoons (pU_{s(p)})^\Xi = \{(pU_{s(p)})^{\alpha^\xi} \mid \xi \in \Xi\}.$$

Furthermore, the initial node satisfies $\Omega_1 = \Omega_0$, $\Xi = \varnothing$, *and terminal nodes of the form* (8) *satisfy the conditions* $\Phi_5(\Omega_1, \mathscr{M})$, $\Phi_6(\Omega_1, \mathscr{M})$ *for* $\mathscr{M} = \Xi + \{s(p)\}$ *and also the condition*

$$\Phi_{13}(a_p, \Omega_1) \rightleftharpoons (\forall \omega_1, \omega_2 \in \Omega_1)(a_p N_0)^{\alpha^{\omega_1}} \cap (a_p N_0)^{\alpha^{\omega_2}} \neq \{0\}.$$

In this section we shall establish that sets of the form $(aN_0)^{\Omega_0}$ are provable (and therefore, establish Theorem 1), assuming that the Main Lemma is true. To achieve this, it suffices to construct, for every special set $\mathscr{C}_1$ of the form (8) satisfying Φ_{13}, Φ_5, Φ_6, a special inference with the initial node $\mathscr{C}_1$ such that its terminal nodes of the form (1) satisfy Φ_4, Φ_5, Φ_6 (and then to apply Lemma 5.2).

We use induction on the degree of a. If $a = 1$, then $(aN_0)^{\Omega_0} = N_0^{\Omega_0} \subseteq \mathscr{T}_0$ and the set $(aN_0)^{\Omega_0}$ is provable by the corollary to Lemma 4.3. Let p be an irreducible factor of a with the maximal value of $|s(p)|$. For $a = a_p$ we have that Φ_{13} and Φ_4 are equivalent, and there is nothing to prove.

Let $a \neq a_p$. Set $a_1 \rightleftharpoons a/a_p$.

Consider an a-special set $\mathscr{C} = \mathscr{A}(\Omega_1) \cup \mathscr{B}(\Xi)$ of the form (8), satisfying Φ_{13}, Φ_5, Φ_6. By part 1 of the proof of Lemma 5.2

$$(a_p N_0)^{\Omega_1} \rightleftharpoons b_1 \cdot N_0^{\Omega_1}, \qquad b_1 = \bigvee_{\omega \in \Omega_1} a_p^{\alpha^\omega}. \qquad (9)$$

Therefore, the set $\mathscr{C}_1$ can be written in the form

$$\mathscr{C}_1 = b_1 \cdot ((a_1 N_0)^{\Omega_1}) \cup \mathscr{B}_1,$$
$$\mathscr{B}_1 \rightleftharpoons \{q_M U_M \mid M \in \mathscr{M}_1\},$$
$$\mathscr{M}_1 \rightleftharpoons \Xi + \{s(p)\}, \qquad q_{\xi + s(p)} \rightleftharpoons p^{\alpha^\xi}.$$

Let

$$c_1 \rightleftharpoons \bigvee_{\substack{\omega_1, \omega_2 \in \Omega_1 \\ \omega_1 \neq \omega_2}} (a_p^{\alpha^{\omega_1}} \wedge a_p^{\alpha^{\omega_2}}), \qquad b_2 \rightleftharpoons b_1 / c_1.$$

The conditions

$$\Phi_5(\Omega_1, \mathscr{M}_1), \qquad \Phi_{10}(\Omega_1, \{a^{\alpha^\omega}\}, c_1), \qquad \Phi_{11}(\Omega_1, c_1)$$

are obviously satisfied, and

$$c_1^{-1} \mathscr{C}_1 = b_2 \cdot (a_1 N_0)^{\Omega_1} \cup \mathscr{B}_1 \qquad (10)$$

(we use (5) and (6)). By the induction assumption, there exists a special inference $\mathscr{D}$ with the initial node $(a_1 N_0)^{\Omega_1}$ such that its terminal nodes are a-special sets of the form (1) satisfying Φ_4, Φ_5, Φ_6. We shall prove that replacing each node

$$\mathscr{C} = \mathscr{A}(\Omega, \{a_\omega\}) \cup \mathscr{B}(\mathscr{M}, \{q_M\}) \qquad (11)$$

of the inference $\mathscr{D}$ by

$$\mathscr{C}' = b_2 \mathscr{A}(\Omega, \{a_\omega\}) \cup \mathscr{B}(\mathscr{M}, \{q_M\}) \cup \mathscr{B}_1, \qquad (12)$$

we obtain a special inference $\mathscr{D}'$ and verify the conditions Φ_4, Φ_5, Φ_6 for the finite nodes of this inference.

LEMMA 6.2. *The set Ω_1 and element b_2 satisfy the condition*

$$(\forall I \in J(b_2)) |\{\omega_1 \in \Omega_1 \mid I \cap (\omega_1 + I_0) \neq \varnothing\}| \leq 1.$$

PROOF. If an irreducible element q divides b_2 and

$$s(q) \cap (\omega_i + I_0) \neq \varnothing, \qquad \omega_i \in \Omega_1, \quad i = 1, 2, \quad \omega_1 \neq \omega_2,$$

then

$$q \mid b_2, \qquad\qquad q^{-1} \notin N_0^{\alpha^{\omega_i}},$$
$$q \mid (b_1)_{J(b_1) \setminus J(N_0^{\alpha^{\omega_i}})} = a_p^{\alpha^{w_i}}, \qquad i = 1, 2,$$
$$q \mid (a_p^{\alpha^{\omega_1}} \wedge a_p^{\alpha^{\omega_2}}), \qquad\qquad q \mid c_1, \quad q^2 / b_1 = b_2 \cdot c_1,$$

But b_1 is square free by (7) and (9). The resulting contradiction proves Lemma 6.2.

Next we prove that $\mathscr{D}'$ is a special inference. Proceed by induction on the inference. The initial node (10) of $\mathscr{D}'$ is obviously a-special. Let $\mathscr{C}$ and $\mathscr{C}'$ be given by (11) and (12), and assume that $\mathscr{C}$ has already been proved to be a-special. First, we consider the case of special rule I being applied to a node $\mathscr{C}$ of $\mathscr{D}$.

Let q be an irreducible polynomial satisfying Φ_7, Φ_8, Φ_9. From $\Phi_7(\Omega, \{a_\omega\}, q)$ we have

$$(\exists \omega_1, \omega_2 \in \Omega)(\omega_1 \neq \omega_2, s(q) \cap (\omega_i + I_0) \neq \varnothing, i = 1, 2).$$

Hence, by Lemma 6.2 and condition $\Phi_5(\Omega_1, \mathscr{M}_1)$

$$\{s(q)\} \vdash J(b_2), \qquad \{s(q)\} \vdash \mathscr{M}_1. \tag{13}$$

This implies, in particular, that $q \nmid b_2$; hence,

$$\Omega^+(\mathscr{C}, q) = \{\omega \in \Omega \mid q \nmid a_\omega\} = \{\omega \in \Omega \mid q \nmid b_2 a_\omega\} = \Omega^+(\mathscr{C}', q), \tag{14}$$

$$\Omega^-(\mathscr{C}, q) \rightleftharpoons \{\omega \in \Omega \mid (q \vee a_\omega) a_\omega^{-1} \in N_0^{\alpha^\omega}\}$$
$$= \{\omega \in \Omega \mid (q \vee a'_\omega)(a'_\omega)^{-1} \in N_0^{\alpha^\omega}\} = \Omega^-(\mathscr{C}', q), \tag{15}$$

where $a'_\omega = h(b_2 a_\omega N_0^{\alpha^\omega})$. Now $\Phi_7(\mathscr{C}', q)$ follows from (14), (15), while $\Phi_8(\mathscr{C}', q)$ and $\Phi_9(\mathscr{C}', q)$ follow from (13) and from $\Phi_8(\mathscr{C}, q)$, $\Phi_9(\mathscr{C}, q)$. Finally, (2), (3), (13), (14), (15), and the equalities

$$F^-(\mathscr{C}', q) = \{b_2 a_\omega N_0^{\alpha^\omega} \mid \omega \in \Omega^-(\mathscr{C}', q)\} \cup \mathscr{B}(\mathscr{M}, \{q_M\}) \cup \mathscr{B}_1,$$
$$F^+(\mathscr{C}', q) = \{b_2 a_\omega N_0^{\alpha^\omega} \mid \omega \in \Omega^+(\mathscr{C}', q)\} \cup \{q U_{s(q)}\}$$
$$\cup \mathscr{B}(\mathscr{M}, \{q_M\}) \cup \mathscr{B}_1$$

yield $F^+(\mathscr{C}', q) = (F^+(\mathscr{C}, q))'$, $F^-(\mathscr{C}', q) = (F^-(\mathscr{C}, q))'$.

Next we consider the case when special rule III is applied to a node $\mathscr{C}$ of the inference $\mathscr{D}$ of the form (11). Let $c \in [\mathbb{Z}]$ be a polynomial satisfying Φ_{10} and Φ_{11}. Condition $\Phi_5(\mathscr{C}')$ follows from $\Phi_5(\mathscr{C})$ and $\Phi_5(\mathscr{C}_1)$ (recall that $c_1^{-1}\mathscr{C}_1$ is an initial node of the inference $\mathscr{D}'$ of the form (10)). Definition (11), (12) immediately show that Φ_{10} and Φ_{11} are carried over from $\mathscr{C}$ to $\mathscr{C}'$. The equality $c^{-1}\mathscr{C}' = (c^{-1}\mathscr{C})'$ follows from (5) and (6).

Thus, we have proved that $\mathscr{D}'$ is a special inference. Let us prove that if a set $\mathscr{C}$ of the form (11) satisfies Φ_4, Φ_5, and Φ_6, then $\mathscr{C}'$ defined by (12) also satisfies these conditions.

Indeed, $\Phi_5(\mathscr{C}')$ has been proved, $\Phi_4(\mathscr{C}')$ follows from $\Phi_4(\mathscr{C})$ since the multiplication by b_2 is an automorphism of the k-space $(\mathbb{Z})$. We must prove $\Phi_6(\mathscr{C}')$. Let $\omega_0 \in \Omega$. If $\{\omega_0 + I_0\} \wedge \mathscr{M} = \{\varnothing\}$, then $\Phi_6(\mathscr{C}', \omega_0)$ follows from $\Phi_6(\mathscr{C}_1, \omega_0)$. By definition,

$$\Phi_6(\mathscr{C}, \omega) \rightleftharpoons \Phi_6(\mathscr{M}, \omega) \rightleftharpoons (\exists \nu \in \mathbb{Z})\{\mathbb{Z}_-^\nu, \mathbb{Z}_+^\nu\} \succcurlyeq (\omega + \{\mathbb{Z}_-^\lambda, \mathbb{Z}_+^\mu\}) \cup \mathscr{M}.$$

Let $\{\omega_0 + I_0\} \wedge \mathscr{M} = \{\varnothing\}$. Following the proof of Lemma 5.2, from sets of the form $\mathscr{N}_{\rho,\sigma} = \{\mathbb{Z}_-^\rho, \mathbb{Z}_+^\sigma\}$, $\rho \leq \sigma$ satisfying

$$\mathbb{N}_{\rho,\sigma} \succcurlyeq (\omega_0 + \{\mathbb{Z}_-^\lambda, \mathbb{Z}_+^\mu\}) \cup \mathscr{M} \tag{16}$$

we choose a minimal one with respect to the order relation $\preceq$. Sets satisfying (16) exist by $\Phi_6(\mathscr{M}, \omega_0)$. Since $M \cap (\omega_0 + I_0) \neq \varnothing$ for some $M \in \mathscr{M}$ we have $\mathscr{N}_{\rho_0, \sigma_0} \succneqq \omega_0 + \{\mathbb{Z}_-^\lambda, \mathbb{Z}_+^\mu\}$, and therefore, one of the two inequalities: $\rho_0 > \omega_0 + \lambda$, $\sigma_0 < \omega_0 + \mu$, is true. The arguments for the two cases being symmetrical, we assume the first inequality holds. Then the minimality of the chosen set implies $\mu(M_0) = \rho_0$ for some $M_0 \in \mathscr{M}$ and $M \subseteq \mathbb{Z}_-^{\rho_0-1} \cup \mathbb{Z}_+^{\sigma_0}$ for all $M \in \mathscr{M} \setminus \{M_0\}$. Let $\nu = \rho_0$. Then

$$\{\mathbb{Z}_-^\nu, \mathbb{Z}_+^\nu\} \succcurlyeq (\omega_0 + \{\mathbb{Z}_-^\lambda, \mathbb{Z}_+^\mu\}) \cup \mathscr{M}.$$

If $\{M_1\} \not\preceq \{\mathbb{Z}_-^\nu, \mathbb{Z}_+^\nu\}$ for some $M_1 \in \mathscr{M}_1$, then $\rho_0 - 1 \in M_1$. Condition $\Phi_{12}(\Omega_1, \Omega, \mathscr{M})$ is true by Lemma 6.1 and implies the existence of $\omega_1 \in \Omega_1 \setminus \Omega$ such that $\omega_1 + I_0 \supseteq M_0$. We have

$$\rho_0 - 1 \in M_0 \cap M_1 \cap (\omega_0 + I_0) \subseteq M_1 \cap (\omega_1 + I_0) \cap (\omega_0 + I_0), \qquad \omega_0, \omega_1 \in M_1 - I_0.$$

From $\omega_0 \in \Omega$ it follows that $\omega_0 \neq \omega_1$, and this contradicts $\Phi_5(\mathscr{C}_1)$. Hence, $\{\mathbb{Z}_-^\nu, \mathbb{Z}_+^\nu\} \succcurlyeq \mathscr{M}_1$.

Therefore, the set $\mathscr{C}'$ satisfies Φ_1–Φ_6 and consequently is provable (by Lemma 5.2). Hence, the set $(aN_0)^{\Omega_0}$ is provable.

It remains to prove the Main Lemma.

§7. Proof of the Main Lemma

Let p be an irreducible factor of an element a such that the value of $|s(p)|$ is maximal. In this section we consider only sets of basic modules of the form

$$\mathscr{C}(\Omega, \Xi) = \mathscr{A}(\Omega) \cup \mathscr{B}(\Xi), \qquad \Omega, \Xi \subseteq \mathbb{Z}, \ |\Omega|, |\Xi| < \infty,$$
$$\mathscr{A}(\Omega) \rightleftharpoons (aN_0)^\Omega, \qquad \mathscr{B}(\Xi) \rightleftharpoons (pU_{s(p)})^\Xi = \{(pU_{s(p)})^{\alpha^\xi} \mid \xi \in \Xi\} \tag{1}$$

and apply to them only special rule I. We also assume that for all germs

$$\mathscr{C} \begin{array}{c} \longleftarrow F^+(\mathscr{C}, q) \\ \longleftarrow F^-(\mathscr{C}, q) \end{array}$$

$q \in \overline{p}$. This rule applied to a set of the form (1) will again yield sets of the form (1). This enables us, instead of sets of basic modules of the form (1), to confine our attention to pairs (Ω, Ξ) of finite sets of integer numbers.

We shall reformulate the conditions Φ_i that will be needed later in terms of sets Ω, Ξ.

The conditions Φ_1, Φ_3 are true for any set of the form (1).

Define
$$I \rightleftharpoons \{i \in \mathbb{Z} \mid p^{\alpha^i} \notin N_0\} = \{i \in \mathbb{Z} \mid (p^{\alpha^i})^{-1} \notin N_0\}$$
$$= \{i \in \mathbb{Z} \mid (i + s(p)) \cap I_0 \neq \varnothing\} = I_0 - s(p).$$

It is rather easy to see that for a set of the form (1), Φ_5 can be rewritten as
$$\Psi_5(\Omega, \Xi) \rightleftharpoons (\forall \xi \in \Xi) \mid \Omega \cap (\xi - I) \mid \leq 1.$$

We replace Φ_6 by a stronger condition:
$$\Phi_6'(\nu, \Omega, \Xi) \rightleftharpoons (\forall \omega \in \Omega) \, \Xi + \{s(p)\} \preceq \{\mathbb{Z}_-^{\omega+\nu}, \mathbb{Z}_+^{\omega+\nu}\},$$

where ν is a number from $\langle \lambda, \mu+1 \rangle$, so that $\{\mathbb{Z}_-^\lambda, \mathbb{Z}_+^\mu\} \preceq \{\mathbb{Z}_-^\nu, \mathbb{Z}_+^\nu\}$. The condition Φ_6' is appropriately written in the form
$$\Psi_6(\nu, \Omega, \Xi) \rightleftharpoons \Xi - \Omega + \{s(p)\} \preceq \{\mathbb{Z}_-^\nu, \mathbb{Z}_+^\nu\}.$$

We define an integer-valued function f on the interval I as follows:
$$f(i) = 1 \Leftrightarrow p^{\alpha^i} \mid a, \qquad f(i) = 0 \Leftrightarrow p^{\alpha^i} \nmid a.$$

Since $J(a_p) \subseteq J(a) \preceq \{I_0\}$, f satisfies
$$f(i) = 1 \implies i + s(p) \subseteq I_0. \tag{2}$$

Condition Φ_2 for a set of the form (2) can be now written as
$$\Psi_2(\Omega, \Xi) \rightleftharpoons (\forall i \in (\Xi - \Omega) \cap I) \, f(i) = 0.$$

Let $q = p^{\alpha^x}$ ($x \in \mathbb{Z}$) be a polynomial defining a special germ of the form I with initial node being a set $\mathscr{E}(\Omega, \Xi)$ of the form (1). Condition Φ_7 becomes
$$\Psi_7(\Omega, x) \rightleftharpoons (\exists \omega_1, \omega_2 \in \Omega \cap (x - I)) \, f(x - \omega_1) > f(x - \omega_2).$$

Condition Φ_8 is automatically true. We shall show that from $\Psi_2(\Omega, \Xi)$ and $\Psi_7(\Omega, x)$ follows $\Phi_9(\Xi + \{s(p)\}, p^{\alpha^x})$. Indeed, if the last condition does not hold, then $x \in \Xi$. By $\Psi_7(\Omega, x)$ there exists a number $\omega_1 \in \Omega \cap (x - I)$ such that $f(x - \omega_1) = 1$, which contradicts $\Psi_2(\Omega, \Xi)$.

In this section we define $\mathbb{Z}$-inference as a tree with nodes corresponding to pairs (Ω, Ξ) of finite sets of integers satisfying $\Psi_2(\Omega, \Xi)$ and with germs of the form
$$(\Omega, \Xi) \diagup\diagdown \begin{array}{c} (\Omega^+(x), \Xi \cup \{x\}) \\ (\Omega^-(x), \Xi) \end{array}$$
where
$$\Omega^+(x) \rightleftharpoons \{\omega \in \Omega \mid f(x - \omega) \neq 1\},$$
$$\Omega^-(x) \rightleftharpoons \{\omega \in \Omega \mid f(x - \omega) \neq 0\}$$
such that each germ satisfies $\Psi_7(\Omega, x)$.

The foregoing formulas imply that if each pair (Ω, Ξ) is replaced by the set $\mathscr{C}(\Omega, \Xi)$ of the form (1), then $\mathbb{Z}$-inference becomes a special inference with all germs belonging to type I.

Denote by $\Pi = \Pi(f)$ the following set of intervals:

$$\Pi \rightleftharpoons \Pi(f) \rightleftharpoons \{\langle \gamma, \delta \rangle \mid \gamma, \delta \in \mathbb{Z}, \gamma < \delta,$$
$$(\forall x \in (\gamma + I) \cap (\delta + I)) \, f(x - \gamma) = f(x - \delta)\}.$$

By 4.1.9,

$$\Pi(f) = \{\langle \gamma, \delta \rangle \mid \langle \gamma, \delta \rangle \neq \varnothing, \, (a_p N_0)^{\alpha^\gamma} \cap (a_p N_0)^{\alpha^\delta} \neq \{0\}\}.$$

In these notations, Φ_{13} becomes

$$\Psi_{13}(\Omega) \rightleftharpoons (\forall \omega_1, \omega_2 \in \Omega) \, (\omega_1 < \omega_2 \implies \langle \omega_1, \omega_2 \rangle \in \Pi(f)).$$

Therefore, to prove the Main Lemma it suffices to prove the following statement.

LEMMA 7.1. *There exists a number* $\nu \in \langle \lambda, \mu+1 \rangle$ *such that for any finite set* $\Omega_0 \subseteq \mathbb{Z}$ *there exists a* $\mathbb{Z}$*-inference* $\mathscr{D}$ *with the initial node* $(\Omega_0, \varnothing)$ *and with every terminal node* (Ω, Ξ) *satisfying* $\Psi_5(\Omega, \Xi)$, $\Psi_6(\nu, \Omega, \Xi)$, $\Psi_{13}(\Omega)$.

Fix a number $\nu \in \langle \lambda, \mu+1 \rangle$. Assume that there exists a function that assigns to every finite set Ω of integers not satisfying $\Psi_{13}(\Omega)$ an integer $x(\Omega)$, in such a way that conditions $\Psi_7(\Omega, x(\Omega))$, $\Psi_{14}(\Omega, x(\Omega))$, $\Psi_{15}(\nu, \Omega, x(\Omega))$ hold. Here Ψ_{14}, Ψ_{15} are defined as follows:

$$\Psi_{14}(\Omega, x) \rightleftharpoons (\forall \omega_1, \omega_2 \in \Omega \cap (x - I))(\langle \omega_1, \omega_2 \rangle \in \Pi \implies f(x - \omega_1) = 1),$$
$$\Psi_{15}(\nu, \Omega, x) \rightleftharpoons (\forall \omega \in \Omega \cap (x - I))$$
$$(f(x - \omega) = 0 \implies x - \omega + \{s(p)\} \preceq \{\mathbb{Z}_-^\nu, \mathbb{Z}_+^\nu\}).$$

If such a function $x(\Omega)$ exists, then a $\mathbb{Z}$-inference $\mathscr{D}$, whose existence is guaranteed by Lemma 7.1, can be constructed by induction as follows.

Let $\mathscr{D}_0$ be the $\mathbb{Z}$-inference with a single node $(\Omega_0, \varnothing)$. Suppose that a $\mathbb{Z}$-inference $\mathscr{D}_k$ is already constructed, and has a terminal node (Ω_1, Ξ_1) not satisfying $\Psi_{13}(\Omega_1)$. We extend $\mathscr{D}_k$ to a $\mathbb{Z}$-inference $\mathscr{D}_{k+1}$ by adding the germ

$$(\Omega_1, \Xi_1) \diagup^{(\Omega_1^+(x(\Omega)), \, \Xi_1 \cup \{x(\Omega_1)\})}_{(\Omega_1^-(x(\Omega_1)), \, \Xi_1)}$$

The condition $\Psi_7(\Omega_1, x(\Omega_1))$ holds by assumption, so that $\mathscr{D}_{k+1}$ indeed is a $\mathbb{Z}$-inference. We proceed with this construction until all terminal nodes of a $\mathbb{Z}$-inference $\mathscr{D}_k$ satisfy Ψ_{13}. Then we stop and define $\mathscr{D} \rightleftharpoons \mathscr{D}_k$. By Ψ_7 the sets Ω^+, Ω^- are proper subsets of Ω, and therefore, the process will terminate after a finite number of steps.

We prove that all terminal nodes of the constructed $\mathbb{Z}$-inference $\mathscr{D}$ satisfy Ψ_5, Ψ_6. The condition Ψ_6 is true for the initial node and is carried over to subsets; hence, it is enough to establish the implication

$$\Psi_6(\Omega, \Xi) \implies \Psi_6(\Omega^+(x(\Omega)), \Xi \cup \{x(\Omega)\}).$$

We have

$$\Xi \cup \{x(\Omega)\} - \Omega^+(x(\Omega)) + \{s(p)\}$$
$$\subseteq (\Xi - \Omega + \{s(p)\}) \cup (x(\Omega) - \Omega^+(x(\Omega)) + \{s(p)\})$$

and it suffices to verify that

$$x(\Omega) - \omega + \{s(p)\} \preceq \{\mathbb{Z}_-^\nu, \mathbb{Z}_+^\nu\} \tag{3}$$

for any $\omega \in \Omega^+(x(\Omega)) = \{\omega \in \Omega \mid f(x(\Omega) - \omega) \neq 1\}$. Let $\omega \in \Omega^+(x(\Omega))$ be a number for which (3) is not true. Then

$$x(\Omega) - \omega + \{s(p)\} \not\preceq \{\mathbb{Z}_-^\lambda, \mathbb{Z}_+^\mu\},$$

and hence, $(x(\Omega) - \omega + s(p)) \cap I_0 \neq \varnothing$, $\omega \in x(\Omega) - I$. Therefore, the value of $f(x(\Omega) - \omega)$ is defined and equals zero by the definition of Ω^+. We have arrived at the contradiction to condition $\Psi_{15}(\nu, \Omega, x(\Omega))$.

Let (Ω_1, Ξ_1) be a terminal node of $\mathbb{Z}$-inference $\mathscr{D}$. Consider the linearly ordered set of nodes that precede the node (Ω_1, Ξ_1). Let

$$\ldots, (\Omega_k, \Xi_k), \ldots, (\Omega_2, \Xi_2), (\Omega_1, \Xi_1)$$

be all such nodes ordered in a natural way. Then the following inclusions hold:

$$\cdots \supset \Omega_k \supset \cdots \supset \Omega_2 \supset \Omega_1, \qquad \cdots \subseteq \Xi_k \subseteq \cdots \subseteq \Xi_2 \subseteq \Xi_1.$$

Let $\xi \in \Xi_1$. Then

$$\xi = x(\Omega_k), \qquad \Omega_{k-1} = \Omega_k^+(\xi)$$

for some k. If $|(\xi - \Omega_1) \cap I| > 1$, then there exists a nonempty segment $\langle \omega_1, \omega_2 \rangle$ such that

$$\omega_1, \omega_2 \in \Omega_1 \cap (\xi - I) \subseteq \Omega_k \cap (\xi - I).$$

By $\Psi_{13}(\Omega_1)$ we have $\langle \omega_1, \omega_2 \rangle \in \Pi$, and by $\Psi_{14}(\Omega_K, \xi)$ we have $f(\xi - \omega_1) = f(\xi - \omega_2) = 1$, so that

$$\omega_1, \omega_2 \notin \Omega_{k-1} \supseteq \Omega_1.$$

The resulting contradiction proves that $\Psi_5(\Omega_1, \Xi_1)$ is satisfied.

Therefore, it suffices to prove the existence of an integer-valued function $x(\Omega)$ defined on the set

$$\{\Omega \subseteq \mathbb{Z} \mid |\Omega| < \infty, \neg\Psi_{13}(\Omega)\}$$

such that for any value of the argument the conditions $\Psi_7(\Omega, x(\Omega))$, $\Psi_{14}(\Omega, x(\Omega))$, $\Psi_{15}(\nu, \Omega, x(\Omega))$ are satisfied. Suppose that there exists a set Ω for which there is no number x satisfying simultaneously the conditions $\Psi_7(\Omega, x(\Omega))$, $\Psi_{14}(\Omega, x(\Omega))$, $\Psi_{15}(\nu, \Omega, x(\Omega))$. Then the following condition holds:

$$\Psi_{16}(\varphi, \Omega) \rightleftharpoons (\forall x \in \mathbb{Z})(\neg\Psi_7(\Omega, x) \vee \neg\Psi_{14}(\Omega, x) \vee \neg\Psi_{15}(\nu, \Omega, x)).$$

Hence, to prove the Main Lemma it suffices to find a number $\nu \in \langle \lambda, \mu + 1 \rangle$ such that for every set Ω of integers $\Psi_{16}(\nu, \Omega)$ implies $\Psi_{13}(\Omega)$.

Introduce the following notation:
$$A_1 : A_2 \rightleftharpoons \{x \in \mathbb{Z} \mid x - A_2 \subseteq A_1\}, \qquad A_1, A_2 \subseteq \mathbb{Z}.$$
Given an integer-valued function g on an interval J, define
$$g_K(x) \rightleftharpoons \sum_{i \in K} g(x-i), \qquad x \in J : K, \ K \subseteq \mathbb{Z};$$
$$\Delta g(x) \rightleftharpoons g(x-1) - g(x), \qquad x \in J \cap (J+1);$$
$$\tilde{g}(x) \rightleftharpoons g(-x), \qquad x \in -J;$$
$$\Pi(g) \rightleftharpoons \{\langle \gamma, \delta \rangle \mid \gamma, \delta \in \mathbb{Z}, \ \gamma < \delta, \ (\forall x \in J : \{\gamma, \delta\}) \Delta(g_{\langle \gamma, \delta \rangle})(x) = 0\}.$$

LEMMA 7.2. *In the above notation, the following relations hold:*

(a) $(-A_1) : (A_2) = -(A_1 : (-A_2))$;
(b) $(A_1 : A_2) : A_3 = A_1 : (A_2 + A_3)$;
(c) $A_1 : A_2 = \bigcap_{a_2 \in A_2}(A_1 + a_2)$;
(d) *For intervals* K, L *with* $K : L \neq \varnothing$, *we have* $\lambda(K : L) = \lambda(K) + \mu(L) - 1$, $\mu(K : L) = \mu(K) + \lambda(L)$;
(e) $(g_K)_L(x) = (g_L)_K(x)$, $K, L \subseteq \mathbb{Z}$, $x \in J : (K + L)$;
(f) *For an interval* K, $(\Delta g)_K(x) = \Delta(g_K)(x) = g(x - \mu(K)) - g(x - \lambda(K))$, $x \in J : \{\lambda(K), \mu(K)\}$;
(g) $(\tilde{g})_K(x) = \widetilde{g_{-K}}(x), x \in (-J) : K, \ K \subseteq \mathbb{Z}$;
(h) $\Delta(\tilde{g})(x) = -\widetilde{\Delta g}(x-1)$, $x \in (-J) \cap (-J+1)$;
(i) $\tilde{\tilde{g}}(x) = g(x)$, $x \in J$;

and the domains in the left- and right-hand sides in each of the equations (e)–(f) *coincide.*

PROOF. All the relations can be verified directly. Let us prove, for example, the most difficult one, namely (f). Let $x \in J : \{\lambda(K), \mu(K)\}$. Then
$$(\Delta g)_K(x) = \sum_{i \in K}(g(x-i-1) - g(x-i)) = g(x - \mu(K)) - g(x - \lambda(K)),$$
$$\Delta(g_K)(x) = \sum_{i \in K+1}(g(x-i) - \sum_{i \in K} g(x-i) = g(x - \mu(K)) - g(x - \lambda(K)).$$
Other relations are proved similarly.

LEMMA 7.3. *In the above notation, let* K, L *be nonempty intervals, let* γ *and* δ *be integers satisfying*
$$\mu(J) + \lambda(L) + \lambda(K) \geq \delta + \lambda(K) \geq \gamma + \lambda(L) \geq \lambda(J) + \mu(K) + \mu(L), \qquad (4)$$
$$(\forall x \in A \rightleftharpoons \langle \lambda(J) + \mu(K), \gamma \rangle) \Delta g_K(x) \geq 0,$$
$$(\forall x \in B \rightleftharpoons \langle \lambda(J) + \mu(L), \delta \rangle) \Delta g_L(x) \leq 0. \qquad (5)$$
Then $(\forall x \in A) \Delta g_K(x) = 0$. *Furthermore, if*
$$\delta + \lambda(K) = \gamma + \lambda(L), \qquad (6)$$
then $(\forall x \in B) \Delta g_L(x) = 0$.

PROOF. Inequalities (4) yield the following two statements:

1. $A \subseteq J: \{\lambda(K), \mu(K)\}$, $B \subseteq J: \{\lambda(L), \mu(L)\}$, and hence, the values of $\Delta g_K(x)$, $\Delta g_i(x)$ in (5) are defined.

2. Denote the interval $A: L = \langle \lambda(J) + \mu(L) + \mu(K) - 1, \gamma + \lambda(L) \rangle$ by C. Then C is nonempty and $C \rightleftharpoons A: L \subseteq B: K = \langle \lambda(J) + \mu(L) + \mu(K) - 1, \delta + \lambda(K) \rangle$. Moreover, in the case (6) we have equality.

Hence, $C - L = A$, $C - K \subseteq B$ ((6) implies $C - K = B$). Let $x \in C$. By Lemma 7.2 (e), (f)

$$0 \leq \sum_{i \in L} \Delta g_K(x - i) = (\Delta g_K)_L(x) = (\Delta g_L)_K(x) = \sum_{j \in K} \Delta g_L(x - j) \leq 0.$$

All the summands in each sum have the same signs due to (5), and so they all equal zero. Therefore,

$$(\forall y \in C - L) \Delta g_K(y) = 0, \qquad (\forall y \in C - K) \Delta g_L(y) = 0.$$

The lemma is proved.

COROLLARY. *Let K, L be nonempty intervals, let $|K| + |L| \leq |J|$, and let*

$$(\forall x \in J: \{\lambda(K), \mu(K)\}) \Delta g_K(x) \geq 0,$$
$$(\forall x \in J: \{\lambda(L), \mu(L)\}) \Delta g_L(x) \leq 0.$$

Then $K, L \in \Pi(g)$.

PROOF. Denote $\gamma \rightleftharpoons \mu(J) + \lambda(K)$, $\delta \rightleftharpoons \mu(J) + \lambda(L)$. Inequalities (4) and (6) hold, and $K, L \in \Pi(g)$ by Lemma 7.3.

Now proceed to the proof of the Main Lemma. Define

$$\begin{aligned}
\Pi_+(f) &\rightleftharpoons \{\langle \rho, \sigma \rangle \neq \varnothing \mid (\forall x \in I: \{\rho, \sigma\}) \Delta f_{\langle \rho, \sigma \rangle}(x) \geq 0\}, \\
\Pi_-(f) &\rightleftharpoons \{\langle \rho, \sigma \rangle \neq \varnothing \mid (\forall x \in I: \{\rho, \sigma\}) \Delta f_{\langle \rho, \sigma \rangle}(x) \leq 0\}, \\
\chi_+ &\rightleftharpoons \min\{\sigma - \rho \mid \langle \rho, \sigma \rangle \in \Pi_+(f)\}, \\
\chi_- &\rightleftharpoons \min\{\sigma - \rho \mid \langle \rho, \sigma \rangle \in \Pi_-(f)\}, \\
\tau_+ &\rightleftharpoons \min(\{\sigma - \rho \mid \langle \rho, \sigma \rangle \in \Pi_+(f) \setminus \Pi(f)\} \cup \{|I_0|\}), \\
\tau_- &\rightleftharpoons \min(\{\sigma - \rho \mid \langle \rho, \sigma \rangle \in \Pi_-(f) \setminus \Pi(f)\} \cup \{|I_0|\}).
\end{aligned} \qquad (7)$$

Obviously, $\Pi(f) = \Pi_+(f) \cap \Pi_-(f)$.

Due to symmetry between Π_+ and Π_-, we can assume without loss of generality that $\tau_+ \geq \tau_-$. For a more rigorous argument we can use the function $\tilde{f}$. It is rather easy to verify that $\Pi_+(f) = \Pi_-(\tilde{f})$, $\Pi_-(f) = \Pi_+(\tilde{f})$, and Lemma 7.2 easily yields that a quintuplet $(I, s(p), f, \nu, \Omega)$ satisfies Ψ_{16} if and only if $(-I, -s(p), \tilde{f}, -\nu + 1, -\Omega)$ satisfies Ψ_{16}.

From now on we assume that $\tau_+ \geq \tau_-$. Due to definitions (7), equality $\tau_+ = \tau_-$ takes place only in the case $\tau_+ = \tau_- = |I_0|$. From the definition of function f and property (2) it follows that if J is an interval and $|J| \geq |I_0|$ then $J \in \Pi(f)$.

Let $\tau_- \neq |I_0|$. Then $\tau_- < \tau_+ \leq |I_0|$, $\chi_+ \leq \tau_+$. Set $J \rightleftharpoons I$, $g \rightleftharpoons f$, $K \rightleftharpoons \langle 0, \chi_+ \rangle$, $L \rightleftharpoons \langle 0, \tau_- \rangle$. If $|K| + |L| \leq |I|$, then the conditions in the Corollary of Lemma 7.3 are satisfied; hence, $L \in \Pi(f)$ and this contradicts the definition of τ_-. Therefore, if $\tau_- \neq |I_0|$, then

$$\chi_+ + \tau_- > |I|. \tag{8}$$

Let $\nu \rightleftharpoons \lambda(I_0) + \tau_+ = \lambda + \tau_+ \in \langle \lambda, \mu + 1 \rangle$. Fix a finite set Ω satisfying $\Phi_{16}(\nu, \Omega)$. Elements of this set will be denoted by the letter ω with subscripts. We shall verify $\Phi_{13}(\Omega)$, i.e., prove that if $\langle \omega_1, \omega_2 \rangle \neq \varnothing$, then $\langle \omega_1, \omega_2 \rangle \in \Pi(f)$. Whenever Lemma 7.3 or its corollary are used during the proof, we assume that $J = I$ and $g = f$.

On the set $\Omega \setminus \{\max \Omega\}$ we consider the function $\omega \longrightarrow \omega^+ \rightleftharpoons \min\{\omega' \in \Omega \mid \omega' > \omega\}$. It suffices to show that $\langle \omega, \omega^+ \rangle \in \Pi(f)$ for every element $\omega \in \Omega \setminus \{\max \Omega\}$. Once this is established, we obtain $f(x - \omega) = f(x - \omega^+) = f(x - \omega^{++}) = \cdots$, and the sequence can be continued as long as the function is well defined.

If $\Phi_{13}(\Omega)$ does not hold, there exists an element

$$\omega_1 \rightleftharpoons \min\{\omega \in \Omega \setminus \{\max \Omega\} \mid \langle \omega, \omega^+ \rangle \notin \Pi(f)\}. \tag{9}$$

Consider the following interval:

$$M \rightleftharpoons (I: \{\omega_1, \omega_1^+\}) \cap (\mathbb{Z}_-^\nu : (\omega_1^+ - s(p))). \tag{10}$$

Let x be the minimal element of this interval satisfying

$$(\exists \omega_2 \in \Omega \cap (x - I)), \quad f(x - \omega_1) > f(x - \omega_2). \tag{11}$$

If $f(x - \omega_1) > f(x - \omega)$, then $\omega \geq \omega_1^+$, since otherwise, $\omega \leq \omega_1$ and $f(x - \omega) = f(x - \omega_1)$ by (9). Let us verify $\Phi_{15}(\nu, \Omega, x)$. If $\omega \in \Omega \cap (x - I)$, $f(x - \omega) = 0$, then $f(x - \omega_1) > f(x - \omega)$, $\omega \geq \omega_1^+$, $x + s(p) \subseteq \omega_1^+ + \mathbb{Z}_-^\nu \subseteq \omega + \mathbb{Z}_-^\nu$. Here we are using that $x \in M \subseteq \mathbb{Z}_-^\nu : (\omega_1^+ - s(p))$. The condition $\Phi_7(\Omega, x)$ holds by (11); hence, $\Phi_{14}(\Omega, x)$ is not satisfied. Let $\langle \omega_3, \omega_4 \rangle \in \Pi(f)$ be an interval such that $f(x - \omega_3) = f(x - \omega_4) = 0$. Then $\omega_1 < \omega_3 < \omega_4$ by (9) and (11), and since $x \in I: \{\omega_1, \omega_4\}$, it follows that

$$y \rightleftharpoons x + \omega_3 - \omega_4 \in I: \{\omega_1, \omega_3\}, \quad z \rightleftharpoons x + \omega_3 - \omega_1 \in I: \{\omega_3, \omega_4\}.$$
$$f(y - \omega_1) = f(z - \omega_4) = f(z - \omega_3) = f(x - \omega_1) > f(x - \omega_4) = f(y - \omega_3).$$

The last inequality contradicts the minimality of x. Therefore, there does not exist $x \in M$ satisfying (11); hence,

$$(\forall x \in M) \, \Delta f_{\langle \omega_1, \omega_1^+ \rangle}(x) \geq 0. \tag{12}$$

Next we prove that $M = I: \{\omega_1, \omega_1^+\}$. Indeed, assuming the contrary,

we have
$$I : \{\omega_1, \omega_1^+\} = (I + \omega_1) \cap (I + \omega_1^+) = \langle \lambda(I) + \omega_1^+, \mu(I) + \omega_1 \rangle$$
$$\not\subseteq \mathbb{Z}_-^\nu : (\omega_1^+ - s(p))$$
$$= \mathbb{Z}_-^\lambda + \tau_+ + \omega_1^+ - \mu(s(p)) + 1$$
$$= \mathbb{Z}_-^{\lambda(I)} + \tau_+ + \omega_1^+$$

so that
$$M = \langle \lambda(I) + \omega_1^+, \lambda(I) + \omega_1^+ + \tau_+ \rangle, \qquad \omega_1^+ - \omega_1 + \tau_+ < |I|. \tag{13}$$

Define
$$K \rightleftharpoons \langle \omega_1, \omega_1^+ \rangle, \quad L \rightleftharpoons \langle 0, \tau_- \rangle, \quad \gamma \rightleftharpoons \lambda(I) + \omega_1^+ + \tau_+, \quad \delta \rightleftharpoons \mu(I),$$

and verify the conditions of Lemma 7.3. The inequalities (4) follow from (13) and from $\tau_+ \geq \tau_-$, and (5) follows from (12) together with the definition of τ_-. By Lemma 7.3 we have
$$(\forall x \in M) \; \Delta f_{\langle \omega_1, \omega_1^+ \rangle}(x) = 0. \tag{14}$$

First, assume that $\tau_+ = |I_0|$. Then
$$M = \lambda(I) + \omega_1^+ + \langle 0, \mu - \lambda \rangle = I_0 + \omega_1^+ - \mu(s(p)) + 1.$$

If $f(x - \omega_1) = 1$, then by (2) $x + s(p) \subseteq I_0 + \omega_1$. Applying (13) together with the equality $I = I_0 - s(p)$ we obtain
$$\omega_1^+ - \omega_1 < |I| - |I_0| = |s(p)| - 1,$$
$$\lambda(M) = \lambda + \omega_1^+ - \mu(s(p)) + 1 < \lambda + \omega_1 + |s(p)| - \mu(s(p))$$
$$= \lambda + \omega_1 - \lambda(s(p)) \leq x \leq \mu + \omega_1 - \mu(s(p))$$
$$< \mu + \omega_1^+ - \mu(s(p)) + 1 = \mu(M),$$

i.e., $x \in M \subseteq I : \{\omega_1, \omega_1^+\}$, $f(x - \omega_1) = f(x - \omega_1^+)$. We have shown that for any y, $f(y) = 1$ implies $f(y - \omega_1^+ + \omega_1) = 1$. This contradicts the finiteness of the interval I, so that
$$|I_0| > \tau_+ > \tau_-. \tag{15}$$

Since $I : \{0, \tau_+\} \subseteq I : \{0, \tau_-\}$ we have
$$(\forall x \in I : \{0, \tau_+\})(f(x - \tau_+) \geq f(x) \geq f(x - \tau_-), \Delta f_{\langle \tau_-, \tau_+ \rangle}(x) \geq 0). \tag{16}$$

Let
$$K \rightleftharpoons \langle \tau_-, \tau_+ \rangle, \quad L \rightleftharpoons \langle \omega_1, \omega_1^+ \rangle, \quad \gamma \rightleftharpoons \mu(I), \quad \delta \rightleftharpoons \lambda(I) + \omega_1^+ + \tau_+.$$

The inequalities in the statement of Lemma 7.3 follow from (13), (8), (14), (15), (16). Applying Lemma 7.2 we get $\Delta f_{\langle \tau_-, \tau_+ \rangle}(x) = 0$ for any $x \in I : \{0, \tau_+\}$; hence, $\langle 0, \tau_+ \rangle \in \Pi(f)$ by (16), contrary to the definition of τ_+ (in (7)) and to the inequalities (14).

The resulting contradiction proves that (13) does not hold, i.e., $\omega_1^+ - \omega_1 + \tau_+ \geq |I|$, $M = I : \{\omega_1, \omega_1^+\}$, and also $\langle \omega_1, \omega_1^+ \rangle \in \Pi_+(f)$. Since $\langle \omega_1, \omega_1^+ \rangle \notin \Pi(f)$ by assumption, it follows that

$$|I_0| > \omega_1^+ - \omega_1 \geq \tau_+ > \tau_-. \qquad (17)$$

Let $\omega_2 > \omega_1$. Assume that

$$\omega_2 - \omega_1 \leq |I|, \qquad \langle \omega_1, \omega_2 \rangle \in \Pi(f). \qquad (18)$$

Consider the segment

$$M_1 \rightleftharpoons (I + \omega_2) \setminus (I + \omega_1^+ + \chi_+).$$

Inequalities (8), (17), and (18) imply $\omega_2 < \omega_1^+ + \chi_+$; hence,

$$M_1 = \langle \lambda(I) + \omega_2, \lambda(I) + \omega_1^+ + \chi_+ \rangle. \qquad (19)$$

Let $x \in M_1$, $f(x - \omega_2) > f(x - \omega_1^+)$. If $f(x - \omega_2) > f(x - \omega)$, then $\omega \geq \omega_1^+$, since otherwise $\omega \leq \omega_1$ and $f(x - \omega) = f(x - \omega_1) = f(x - \omega_2)$. Recalling that $x \notin I + \omega_1^+ + \chi_+$ this yields the inclusion

$$x + s(p) \subseteq \mathbb{Z}_-^\lambda + \omega_1^+ + \chi_+ \subseteq \mathbb{Z}_-^\nu + \omega,$$

i.e., $\Psi_{15}(\nu, \Omega, x)$ is satisfied. The condition $\Psi_7(\Omega, x)$ holds due to the choice of x. Thus, $\Psi_{14}(\Omega, x)$ does not hold. Let $\langle \omega_3, \omega_4 \rangle \in \Pi(f)$ be a segment such that $f(x - \omega_3) = f(x - \omega_4) = 0$. Then

$$1 = f(x - \omega_2) = f(x - \omega_1) \neq f(x - \omega_3),$$
$$\omega_3 \geq \omega_1^+, \qquad \omega_4 \geq \omega_3 + \chi_+ \geq \omega_1^+ + \chi_+,$$
$$x \in (\omega_1^+ + I) \cap (\omega_4 + I) \subseteq \omega_1^+ + \chi_+ + I,$$

and this contradicts $x \in M$. Hence,

$$(\forall x \in M_1) \, \Delta f_{\langle \omega_1^+, \omega_2 \rangle}(x) \leq 0. \qquad (20)$$

The inequalities (17) also imply that $I : \{\omega_1, \omega_1^+\} \subseteq I : \{\omega_1 + \tau_-, \omega_1^+\}$; hence,

$$(\forall x \in I : \{\omega_1, \omega_1^+\})(f(x - \omega_1^+) \geq f(x - \omega_1) \geq f(x - \omega_1 - \tau_-),$$
$$\Delta f_{\langle \omega_1 + \tau_-, \omega_1^+ \rangle}(x) \geq 0). \qquad (21)$$

Let

$$K \rightleftharpoons \langle \omega_1 + \tau_-, \omega_1^+ \rangle, \quad L \rightleftharpoons \langle \omega_1^+, \omega_2 \rangle, \quad \gamma \rightleftharpoons \mu(I) + \omega_1, \quad \delta \rightleftharpoons \lambda(I) + \omega_1^+ + \chi_+.$$

All desired inequalities follow from (8), (17), (18), (20), (21). Applying Lemma 7.3 we obtain

$$(\forall x \in I : \{\omega_1, \omega_1^+\}) \, \Delta f_{\langle \omega_1 + \tau_-, \omega_1^+ \rangle}(x) = 0,$$

and by (21) $\langle \omega_1, \omega_1^+ \rangle \in \Pi(f)$, contradicting the assumption.

Consequently, (18) is false, i.e., if $\langle \omega_1, \omega_2 \rangle \in \Pi(f)$, then

$$\omega_2 - \omega_1 > |I|, \qquad (\omega_2 + I) \cap (\omega_1 + I) = \varnothing. \tag{22}$$

Now let $f(x - \omega_1^+) > f(x - \omega_1)$ for some $x \in I : \{\omega_1, \omega_1^+\}$. If $\omega \in \Omega$, $\omega \leq \omega_1$, then $x + s(p) \subseteq I_0 + \omega_1^+ \subseteq \mathbb{Z}_+^\lambda + \omega_1^+ \subseteq \mathbb{Z}_+^\lambda + \tau_+ + \omega_1 = \mathbb{Z}_+^\nu + \omega_1 \subseteq \mathbb{Z}_+^\nu + \omega$. On the other hand, if $\omega \geq \omega_1^+$, then by (10) and the equality $M = I : \{\omega_1, \omega_1^+\}$ we have $x \in \mathbb{Z}_-^\nu : (\omega_1^+ - s(p))$, i.e.,

$$x + s(p) \subseteq \mathbb{Z}_-^\nu + \omega_1^+ \subseteq \mathbb{Z}_-^\nu + \omega.$$

We have verified the condition $\Psi_{15}(\nu, \Omega, x)$. The condition $\Psi_7(\Omega, x)$ holds due to the choice of x. Let us verify $\Psi_{14}(\Omega, x)$. Let $\langle \omega_3, \omega_4 \rangle \in \Pi(f)$, $\{\omega_3, \omega_4\} \subseteq \Omega \cap (x - I)$. If $\omega_3 < \omega_1$, then

$$x \in (\omega_3 + I) \cap (\omega_1^+ + I) \neq \varnothing, \qquad \omega_1^+ - \omega_3 < |I|.$$

Letting $K \rightleftharpoons \langle \omega_1, \omega_1^+ \rangle$, $L \rightleftharpoons \langle \omega_3, \omega_1 \rangle$, and applying the corollary of Lemma 7.3, we arrive at $\langle \omega_1, \omega_1^+ \rangle \in \Pi(f)$. This is a contradiction. Indeed, if $\omega_3 = \omega_1$, then

$$\langle \omega_1, \omega_4 \rangle \in \Pi(f), \qquad x \in (I + \omega_1) \cap (I + \omega_4),$$

which is impossible by (22). On the other hand, if $\omega_3 \geq \omega_1^+$, then

$$\omega_4 - \omega_1 \geq (\omega_4 - \omega_3) + (\omega_1^+ - \omega_1) \geq \chi_+ + \tau_+ \geq \chi_+ + \tau_- \geq |I|,$$

which contradicts the inclusion $x \in (I + \omega_1) \cap (I + \omega_4)$.

Thus, assuming that the set Ω satisfies $\Psi_{16}(\nu, \Omega)$, but not $\Psi_{13}(\Omega)$, we have arrived at the contradiction. This completes the proof of the Main Lemma and of the theorem.

Chapter 2
Inverting Homomorphisms of Rings

In §1 of the second chapter we describe the construction of localization of an arbitrary ring R with unity with respect to an arbitrary set Σ of rectangular matrices. The main result of §1 is Theorem 4, the statement of which is analogous to Ore's Theorem. As a corollary, we are able to produce a system of quasi-identities that hold if and only if the set of matrices in question is potentially invertible.

The main result of §2 is Theorem 7 stating that the set of all nonzero elements of a 2-FI-ring is potentially invertible. In order to establish this result we introduce a notion of independent set of matrices and prove that all upper-triangular matrices over an arbitrary 2-FI-ring form an independent set. The developed methods allow us to derive a sufficient condition for a set $R\Sigma^{-1}$ to be an n-FI-ring (Theorem 8) and to verify this condition for the set $\Sigma_l(R)$ of all square matrices of order $l > 0$ over an $(n+2l)$-FI-ring (Theorem 9).

We assume all rings to have unity preserved under homomorphisms. If R is an arbitrary ring, $^mR^n$ denotes the set of matrices over R with m rows and n columns. If m or n equals 0, the set by definition consists of a single empty matrix. By $^m0^n$ and 1_n we denote the zero matrix from $^mR^n$ and the unit matrix of order n, respectively. If one of the superscripts in the notation $^mR^n$ is omitted, it means that its value is either arbitrary or uniquely determined from the context. The same is true for notations $^m0^n$, 1_n, and other similar notations. By $\mathbb{N}$ we denote the set of all nonnegative integers.

The set $\mathrm{Mat}(R) = \bigcup_{m,n \in \mathbb{N}} {}^mR^n$ is endowed with partial addition

$$\begin{cases} {}^mR^n \times {}^mR^n \to {}^mR^n, & m, n \in \mathbb{N}, \\ (a,b) \to a+b, \end{cases}$$

and multiplication

$$\begin{cases} {}^mR^n \times {}^nR^k \to {}^mR^k, & m, n, k \in \mathbb{N}, \\ (a,b) \to ab, \end{cases}$$

which turns this set into a preadditive category. Its morphisms are matrices and its objects are nonnegative integers, where the number 0 is the zero object, i.e., $1_0 = {}^0 0^0$.

Together with the category $\mathrm{Mat}(R)$, it is convenient to consider the category $\mathrm{Mat}^2(R)$. Its elements (morphisms) are matrices with fixed partition into blocks, where some blocks might be empty. The operation (addition or multiplication) on two matrices from $\mathrm{Mat}^2(R)$ is defined if and only if their block partitions agree, i.e., the required operations can be performed on the blocks. The result of an operation is the usual sum (or product) of matrices, and its block partition is naturally defined by the partitions of summands (or factors). All computations on block matrices are performed in the category $\mathrm{Mat}^2(R)$ unless indicated otherwise.

§1. Description of the construction of localization

We shall use the notions of Σ-inverting homomorphism, localization, and potentially invertible set, given in the Introduction.

A set Σ of rectangular matrices over a ring R is called *multiplicative* if it satisfies the following conditions:

1. The empty (0×0)-matrix 1_0 belongs to Σ.

2. If $a, b \in \Sigma$, then $\begin{pmatrix} a & c \\ 0 & b \end{pmatrix} \in \Sigma$ for any matrix c of appropriate size.

Given a set Σ, the set

$$\overline{\Sigma} \rightleftharpoons \{1_0\} \cup \left\{ \begin{pmatrix} a_{11} & * & * & \cdots & * \\ 0 & a_{22} & * & \cdots & * \\ 0 & 0 & a_{33} & \cdots & * \\ & & & \ddots & \\ 0 & 0 & 0 & \cdots & a_{nn} \end{pmatrix} \middle| a_{ii} \in \Sigma \right\}$$

is the minimal multiplicative set containing Σ; we say that $\overline{\Sigma}$ is the *multiplicative closure* of Σ. Localizations of a ring R with respect to Σ and $\overline{\Sigma}$ are naturally isomorphic; hence, we can, without loss of generality, consider only multiplicative sets.

In this section we shall construct the localization of a ring R with respect to the multiplicative set Σ. For future applications, it is convenient to construct not the ring $R\Sigma^{-1}$ itself, but the entire set $\mathrm{Mat}(R\Sigma^{-1})$ of rectangular matrices over this ring together with partial operations of addition and multiplication defined on this set. First we shall construct a certain set $M(R, \Sigma)$ with partial operations $\oplus, \odot$, an operation-preserving mapping $\varepsilon : \mathrm{Mat}(R) \to M(R, \Sigma)$, and an equivalence relation θ_Σ stable under these operations. Then we shall establish a correspondence between equivalence classes and matrices over a certain ring such that the restriction of the mapping ε to (1×1)-matrices defines the universal Σ-inverting homomorphism.

The following quadruple will be called an $\mathscr{M}$-*object*:
1. a nonempty set M;
2. a partition $M = \bigcup_{m,n \in \mathbb{N}} {}^m M^n$ into disjoint subsets;
3. a family of mappings

$$\begin{cases} {}^m M^n \times {}^m M^n \to {}^m M^n, & m, n, k, \in \mathbb{N}, \\ (a, b) \to a \oplus b; \end{cases}$$

4. a family of mappings

$$\begin{cases} {}^m M^n \times {}^n M^k \to {}^m M^k, & m, n \in \mathbb{N}, \\ (a, b) \to a \odot b. \end{cases}$$

Our main example of an $\mathscr{M}$-object is the set of all rectangular matrices over a ring R with partial operations $+$ and $\cdot$ defined on it. Similarly, any small preadditive category whose objects are nonnegative integers can be viewed as an $\mathscr{M}$-object. Below we shall define an $\mathscr{M}$-object $M(R, \Sigma)$ and an equivalence relation θ_Σ on it in such a way that equivalence classes will correspond to matrices over the ring $R\Sigma^{-1}$.

Given two $\mathscr{M}$-objects M_1 and M_2, we define $\mathscr{M}$-morphism from M_1 to M_2 as the family of mappings:

$$f = \{{}^m f^n : {}^m M_1^n \to {}^m M_2^n\},$$

such that

$$f(a \oplus b) = f(a) \oplus f(b), \qquad f(c \odot d) = f(c) \odot f(d)$$

whenever the results of $a \oplus b$, $c \odot d$ are defined.

Any ring homomorphism $f: R \to R_1$ induces the $\mathscr{M}$-morphism $\mathrm{Mat}(f): \mathrm{Mat}(R) \to \mathrm{Mat}(R_1)$.

We say that an equivalence relation θ on an $\mathscr{M}$-object M is a *congruence*, if the following conditions are satisfied:
1. $(a\theta b, a \in {}^m M^n) \Rightarrow b \in {}^m M^n$;
2. $(a\theta b, c\theta d) \Rightarrow (a \oplus c)\theta(b \oplus d)$, if $a \oplus c$ is defined;
3. $(a\theta b, c\theta d) \Rightarrow (a \odot c)\theta(b \odot d)$, if $a \odot c$ is defined.

If $f: M_1 \to M_2$ is an $\mathscr{M}$-morphism, then the relation

$$\theta_f : a\theta_f b \iff f(a) = f(b)$$

is clearly a congruence. We call it the *kernel congruence* of a morphism f.

Given a congruence θ on an $\mathscr{M}$-object M, the structure of an $\mathscr{M}$-object can be naturally carried over to the quotient set M/θ.

Let R be an arbitrary ring. We define an *R-object* as an $\mathscr{M}$-object M with a fixed $\mathscr{M}$-morphism $\varepsilon: \mathrm{Mat}(R) \to M$. Given two R-objects (M_1, ε_1),

(M_2, ε_2), we define an R-morphism to be an $\mathcal{M}$-morphism $f: M_1 \to M_2$ such that the following diagram is commutative.

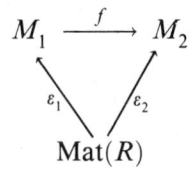

$$\begin{array}{ccc} M_1 & \xrightarrow{f} & M_2 \\ & \varepsilon_1 \searrow \swarrow \varepsilon_2 & \\ & \mathrm{Mat}(R) & \end{array}$$

We say that a congruence θ on an R-object (M, ε) is a *ring congruence* if the quotient object M/θ is a preadditive category (i.e., the operations $\oplus$ and $\odot$ are associative and have respective unit elements, the $\langle {}^m M^n/\theta\,;\oplus\rangle$ are abelian groups and the distributivity identities hold) and the mapping ε/θ in the commutative diagram

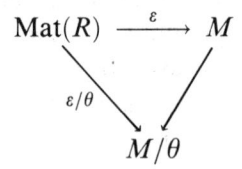

$$\begin{array}{ccc} \mathrm{Mat}(R) & \xrightarrow{\varepsilon} & M \\ & \varepsilon/\theta \searrow \swarrow & \\ & M/\theta & \end{array}$$

is a morphism of preadditive categories (i.e., zero and the unit elements are mapped to zero and to unit elements, respectively).

In particular, this definition implies that the integer 0 is the zero object in M/θ.

PROPOSITION 1. *Let θ be a ring congruence on an R-object (M, ε). Then there is a natural homomorphism of R-objects*

$$\varphi_\theta : (M/\theta, \varepsilon/\theta) \to (\mathrm{Mat}({}^1 M^1/\theta), \mathrm{Mat}({}^1 \varepsilon^1/\theta)).$$

PROOF. Let ${}^m E_{ij}^n \in {}^m R^n$ be a matrix with the (i, j)th element equal to 1 and other elements equal to 0. Let ${}^m e_{ij}^n$ be the image of this matrix under ε/θ. To each element $a \in {}^m M^n/\theta$ we assign the matrix $\varphi_\theta(a) \in {}^m ({}^1 M^1/\theta)^n$ with the (i, j)th element equal to ${}^1 e_{1i}^m a {}^n e_{j1}^1$. It is rather easy to verify that φ_θ is a natural homomorphism of R-objects.

The proposition is proved.

We say that a ring congruence θ on an R-object is the (universal) Σ-inverting congruence, if the ring homomorphism ${}^1\varepsilon^1/\theta : R \to {}^1 M^1/\theta$ is (the universal) Σ-inverting homomorphism. Here Σ is an arbitrary set of matrices over R.

PROPOSITION 2. *Let R be a ring, let Σ be a set of matrices over R, and let (M, ε) be an R-object. Let*

$$\begin{cases} \Sigma \to M, \\ \Sigma \cap {}^m R^n \to {}^n M^m, \\ a \to a^* \end{cases}$$

be a mapping such that M is generated as an R-object by the set $\{a^* | a \in \Sigma\}$ and for any Σ-inverting homomorphism $f: R \to R_1$ the mapping $a^* \to f(a)^{-1}$, $a \in \Sigma$, can be extended to an R-morphism ψ_f from M to the R-object $(\mathrm{Mat}(R_1), \mathrm{Mat}(f))$. Let

$$\theta_\Sigma = \bigcap_f \theta_{\psi_f},$$

where the intersection is taken over all Σ-inverting homomorphisms.

Then θ_Σ is a ring congruence, and the ring homomorphism ${}^1\varepsilon^1/\theta_\Sigma: R \to {}^1M^1/\theta_\Sigma$ is Σ-inverting.

PROOF. For any Σ-inverting homomorphism f the congruence θ_{ψ_f} contains all pairs of the form

$(a \oplus b, b \oplus a), \quad ((a \oplus b) \oplus c, a \oplus (b \oplus c)), \quad (a \oplus \varepsilon(0), a),$
$(a \oplus (\varepsilon(-1) \odot a), \varepsilon(0)), \quad a, b, c \in M,$
$((a \odot b) \odot c, a \odot (b \odot c)), \quad (a \odot \varepsilon(1), a), \quad (\varepsilon(1) \odot a, a), \quad a, b, c \in M,$
$(a \odot (b \oplus c), (a \odot b) \oplus (a \odot c)), \quad a, b, c \in M,$
$((a \oplus b) \odot c, (a \odot c) \oplus (b \odot c)), \quad a, b, c \in M,$
$(a^* \odot \varepsilon(a), \varepsilon(1)), \quad (\varepsilon(a) \odot a^*, \varepsilon(1)), \quad a \in \Sigma.$
(1)

Conversely, any congruence θ containing all pairs of the form (1) is a ring congruence, and hence, defines a homomorphism $f = {}^1\varepsilon^1/\theta: R \to {}^1M^1/\theta$ which is Σ-inverting by Proposition 1.

The mappings $f \to \theta_{\psi_f}$, $\theta \to {}^1\varepsilon^1/\theta$ define a one-to-one correspondence between congruences containing all pairs of the form (1) and classes of isomorphic Σ-inverting homomorphisms $f: R \to R_1$ such that $(\mathrm{Mat}(R_1), \mathrm{Mat}(f))$ is generated by the set $\{f(a)^{-1} | a \in \Sigma\}$ as an R-object. (Two Σ-inverting homomorphisms $f_i: R \to R_i$, $i = 1, 2$, are isomorphic if there exists a ring isomorphism $g: R_1 \to R_2$ such that the following diagram is commutative.

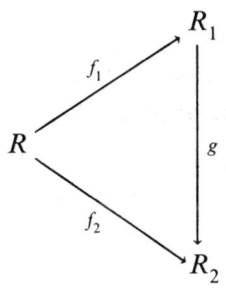

$$\begin{array}{ccc} & & R_1 \\ & \nearrow f_1 & \\ R & & \downarrow g \\ & \searrow f_2 & \\ & & R_2 \end{array}$$

Therefore, the minimal congruence θ_Σ containing all pairs of the form (1) defines the universal Σ-inverting homomorphism.

The proposition is proved.

Now we construct an R-object $(M(R, \Sigma), \varepsilon)$ satisfying the conditions of Proposition 2.

Denote by $M(R)$ the set of rectangular matrices with a fixed block partitioning of the form

$$a = \left(\begin{array}{c|c} a' & \tilde{a} \\ \hline a^0 & {'a} \end{array} \right), \qquad (2)$$

i.e., the set of quadruples of matrices $(a', \tilde{a}, a^0, {'a})$ such that their sizes allow us to form a matrix in (2). We shall view partition (2) as distinguished among all other possible partitions of a. Hence, there are four mappings defined on the set $M(R)$:

$$a \to a', \quad a \to \tilde{a}, \quad a \to a^0, \quad a \to {'a}.$$

Some of the four distinguished blocks of the matrix $a \in M(R)$ may be empty. In this case we shall use shorthand notation, e.g., $(\overline{a^0|})$, $(a'|\tilde{a})$, etc.

Let

$${}^m M^n(R) \rightleftharpoons \{a \in M(R) \mid \tilde{a} \in {}^m R^n\};$$

$$a \oplus b \rightleftharpoons \left(\begin{array}{cc|c} a' & b' & \tilde{a} + \tilde{b} \\ \hline a^0 & 0 & {'a} \\ 0 & b^0 & {'b} \end{array} \right), \qquad a, b \in {}^m M^n(R), \quad m, n \in \mathbb{N};$$

$$a \odot b \rightleftharpoons \left(\begin{array}{cc|c} a' & \tilde{a}b' & \tilde{a}\tilde{b} \\ \hline a^0 & {'a}b' & {'a}\tilde{b} \\ 0 & b^0 & {'b} \end{array} \right), \qquad a \in {}^m M^n(R), \; b \in {}^n M^k(R),\\ m, n, k \in \mathbb{N};$$

$$\varepsilon(a) \rightleftharpoons (\underline{a}), \qquad a \in \mathrm{Mat}(R);$$

$$a^* \rightleftharpoons \left(\begin{array}{c|c} -1 & 0 \\ \hline a & 1 \end{array} \right), \qquad a \in \mathrm{Mat}(R).$$

The first four equalities define the R-object structure on the set $M(R)$.

If $f: R \to R_1$ is a ring homomorphism, the induced mapping $M(f): M(R) \to M(R_1)$ can be regarded as an R-morphism due to the commutativity of the diagram

$$\begin{array}{ccc} M(R) & \xrightarrow{M(f)} & M(R_1) \\ \varepsilon \uparrow & & \uparrow \varepsilon \\ \mathrm{Mat}(R) & \xrightarrow{\mathrm{Mat}(f)} & \mathrm{Mat}(R_1) \end{array}$$

Note also that the mapping $M(f)$ preserves the operation $a \to a^*$.

Let Σ be a multiplicative set of matrices. Then the set

$$M(R, \Sigma) \rightleftharpoons \{a \in M(R) \mid a^0 \in \Sigma\}$$

is closed under the operations $\oplus$ and $\odot$. Hence, on $M(R, \Sigma)$ there is an induced R-object structure, which we also denote by $M(R, \Sigma)$, assuming the mapping ε to be fixed. The obvious identity

$$\left(\begin{array}{c|c} a' & \tilde{a} \\ \hline a^0 & 'a \end{array}\right) = (\lfloor\tilde{a}\rfloor) \oplus \left((\lfloor -a'\rfloor) \odot (\lfloor a^0\rfloor)^* \odot (\lfloor 'a\rfloor)\right)$$

shows that $M(R, \Sigma)$ is generated by the set $\{a^* | a \in \Sigma\}$ as an R-object.

Denote by $\hat{a}$ the matrix

$$\begin{pmatrix} 1 & a' & \tilde{a} \\ 0 & a^0 & 'a \\ 0 & 0 & -1 \end{pmatrix}.$$

In particular, if $p \in {}^m M^0(R)$, $q \in {}^0 M^n(R)$, then

$$\hat{p} = \begin{pmatrix} 1_m & p' \\ 0 & p^0 \end{pmatrix}, \qquad \hat{q} = \begin{pmatrix} q^0 & 'q \\ 0 & -1_n \end{pmatrix}.$$

If $a \in {}^m M^n(R)$ and the matrix a^0 is invertible, then so is $\hat{a}$, and there exists a unique matrix $\omega(a) \in {}^m M^n(R)$ such that

$$\hat{a}^{-1} = \widehat{\omega(a)}.$$

The following identities can be verified directly:

$$\omega(a \oplus b) = \omega(a) \oplus \omega(b),$$
$$\omega(a \odot b) = \omega(a) \odot \omega(b),$$
$$\omega(\lfloor a \rfloor) = (\lfloor a \rfloor), \qquad (3)$$
$$\omega(a^*) = \left(\begin{array}{c|c} a^{-1} & a^{-1} \\ \hline a^{-1} & a^{-1} \end{array}\right).$$

Each of the identities holds if and only if the expression in its left- or right-hand side is well defined. In particular, if Σ consists of invertible matrices and is closed under inversion, then the identities (3) are true for any matrices belonging to $M(R, \Sigma)$. The first three identities imply that in this case the mapping ω is an automorphism of the R-object $M(R, \Sigma)$.

Let $\tilde{\omega}(a) \rightleftharpoons \widetilde{\omega(a)}$. It follows from the definition of ω that $\tilde{\omega}(a) = \tilde{a} - a' \cdot (a^0)^{-1} \cdot 'a$.

Now let $f: R \to R_1$ be a Σ-inverting homomorphism. Denote by Σ_1 the set $f(\Sigma) \cup f(\Sigma)^{-1}$, which is a union of images and of preimages of matrices from Σ. Since Σ_1 consists of invertible matrices and is closed under inversion, it follows that $\tilde{\omega}: M(R_1, \Sigma_1) \to \mathrm{Mat}(R_1)$ is an R_1-morphism and the mapping

$$\psi_f = \tilde{\omega} \circ M(f): M(R, \Sigma) \to \mathrm{Mat}(R_1)$$

is an R-morphism (to an R-object $(\mathrm{Mat}(R_1), \mathrm{Mat}(f)))$. From the fourth identity in (3) we have $\psi_f(a^*) = f(a)^{-1}$.

Therefore, all conditions of Proposition 2 hold, and $\theta_\Sigma = \bigcap_f \theta_{\psi_f}$ is a universal Σ-inverting congruence.

Next we give an explicit description of the congruence θ_Σ.

We say that a set Σ of matrices over a ring R is *saturated* if it contains all matrices $a \in \text{Mat}(R)$ such that the matrix $U_\Sigma(a)$ is invertible over $R\Sigma^{-1}$. (U_Σ stands for universal Σ-inverting homomorphism.) For any Σ there exists a unique saturated set $\overline{\Sigma} \supseteq \Sigma$ such that the universal Σ- and $\overline{\Sigma}$-inverting homomorphisms coincide. Clearly, the universal Σ-inverting congruence θ_Σ on the set $M(R, \Sigma)$ is the restriction of the congruence $\theta_{\overline{\Sigma}}$ (defined on $M(R, \overline{\Sigma})$) to this set. The set $\overline{\Sigma}$ is called the saturation of Σ.

We say that the matrix $c \in M(R, \Sigma)$ is Σ-*incomplete* if it can be written in the form

$$\left(\begin{array}{c|c} c' & \tilde{c} \\ \hline c^0 & 'c \end{array}\right) = \left(\begin{array}{c} a' \\ \hline a^0 \end{array}\right) \cdot \left(\begin{array}{c|c} b^0 & 'b \end{array}\right), \qquad a^0, b^0 \in \overline{\Sigma},$$

where $\overline{\Sigma}$ is the saturation of Σ.

LEMMA 1. *Any Σ-incomplete matrix represents the zero matrix over the ring $R\Sigma^{-1}$.*

PROOF. We can assume that Σ is saturated. It is enough to prove that $\psi_f(c) = 0$ for any Σ-inverting homomorphism f. Therefore, we can assume that Σ consists of invertible matrices and is closed under inversion. Under these assumptions it can immediately be verified that

$$\omega(c) = \left(\begin{array}{c|c} a'_1 & 0 \\ \hline b^0_1 a^0_1 & 'b_1 \end{array}\right),$$

where $a_1 = \omega(a)$, $b_1 = \omega(b)$. Hence, $\tilde{\omega}(c) = 0$.

The lemma is proved.

Given matrices $a, b \in M(R)$, we define

$$\ominus b \rightleftharpoons (\underline{|-1}\,) \odot b, \qquad a \ominus b \rightleftharpoons a \oplus (\ominus b).$$

LEMMA 2. *The following equalities are equivalent in the category* $\text{Mat}^2(R)$:

$$\begin{pmatrix} a' & \tilde{a} \\ a^0 & 'a \end{pmatrix} \cdot \begin{pmatrix} q^0 & 'q \\ 0 & -1 \end{pmatrix} + \begin{pmatrix} 1 & p' \\ 0 & p^0 \end{pmatrix} \cdot \begin{pmatrix} b' & \tilde{b} \\ b^0 & 'b \end{pmatrix} = 0,$$

$$\begin{pmatrix} a' & p' \\ a^0 & p^0 \end{pmatrix} \cdot \begin{pmatrix} q^0 & 'q \\ b^0 & 'b \end{pmatrix} = \begin{pmatrix} -b' & \tilde{a} - \tilde{b} \\ 0 & 'a \end{pmatrix},$$

$$\left(\begin{array}{c|c} a' & \tilde{a} \\ \hline a^0 & 'a \end{array}\right) \ominus \left(\begin{array}{c|c} b' & \tilde{b} \\ \hline b^0 & 'b \end{array}\right) = \left(\begin{array}{c} a' & p' \\ a^0 & p^0 \\ 0 & 1 \end{array}\right) \cdot \left(\begin{array}{cc|c} 1 & q^0 & 'q \\ 0 & b^0 & 'b \end{array}\right).$$

PROOF. We can immediately verify that each of the three equalities above is equivalent to the following equalities:

$$a' \cdot q^0 + p' \cdot b^0 = -b', \qquad a' \cdot {}'q + p' \cdot {}'b = \tilde{a} - \tilde{b},$$
$$a^0 \cdot q^0 + p^0 \cdot b^0 = 0, \qquad a^0 \cdot {}'q + p^0 \cdot {}'b = {}'a.$$

The lemma is proved.

COROLLARY. *If $a, b \in M(R, \Sigma)$, $p \in M^0(R)$, $q \in {}^0 M(R)$, and the equality*

$$a\hat{q} + \hat{p}b = 0 \qquad (4)$$

holds in the category $\mathrm{Mat}^2(R)$, *i.e., the equality*

$$\begin{pmatrix} a' & \tilde{a} \\ a^0 & {}'a \end{pmatrix} \cdot \begin{pmatrix} q^0 & {}'q \\ 0 & -1 \end{pmatrix} + \begin{pmatrix} 1 & p' \\ 0 & p^0 \end{pmatrix} \cdot \begin{pmatrix} b' & \tilde{b} \\ b^0 & {}'b \end{pmatrix} = 0$$

holds, then $a\theta_\Sigma b$.

The assertion follows immediately from Lemmas 1 and 2.

Fix a multiplicative set Σ of matrices over a ring R.

If the conditions of the corollary are satisfied, we shall say that matrix a is left adjacent to b, b is right adjacent to a, and denote this by $a \to b$.

LEMMA 3. *The relation $\to$ is reflexive and transitive. Also, if $a \to b$ and $c \to d$, then $a \oplus c \to b \oplus d$, $a \odot c \to b \odot d$ whenever the corresponding operations are well defined.*

PROOF. Reflexivity follows from $a \cdot (-1) + 1 \cdot a = 0$. If $a\hat{q} + \hat{p}b = 0$, $b\hat{v} + \hat{u}c = 0$ are equalities of the form (4), then the equality $a \cdot (-\widehat{qv}) + (\widehat{pu}) \cdot c = 0$ is also of the form (4). Therefore, $a \to c$.

Let $a\hat{q} + \hat{p}b = 0$, $c\hat{v} + \hat{u}d = 0$. If $a \oplus c$ is defined, then

$$(a \oplus c) \cdot (q \oplus v) + (p \oplus v) \cdot (b \oplus d) = 0.$$

If $a \cdot c$ is defined, then

$$(a \odot c) \cdot \left(\widehat{q \odot \left(\begin{array}{c|c} d' & \tilde{d} \\ \hline v^0 & {}'v \end{array} \right)} \right) + \left(\widehat{\left(\begin{array}{c|c} p' & \tilde{a} \\ \hline a^0 & {}'a \end{array} \right) \odot u} \right) \cdot (b \odot d) = 0.$$

Both equalities can be immediately verified. The lemma is proved.

To perform computations in the universal Σ-inverting ring we need the following

LEMMA 4. *Consider the following pairs of matrices*:

1. $\left(\begin{array}{ccc|c} c'_1 & c'_2 & c'_3 & \tilde{c} \\ \hline c_{11} & c_{12} & c_{13} & 'c_1 \\ 0 & c_{22} & 0 & 0 \\ c_{31} & c_{32} & c_{33} & 'c_3 \end{array}\right)$, $\left(\begin{array}{cc|c} c'_1 & c'_3 & \tilde{c} \\ \hline c_{11} & c_{13} & 'c_1 \\ c_{31} & c_{33} & 'c_3 \end{array}\right)$;

2. $\left(\begin{array}{cc|c} c'_1 & c'_3 & \tilde{c} \\ \hline c_{11} & c_{13} & 'c_1 \\ c_{31} & c_{33} & 'c_3 \end{array}\right)$, $\left(\begin{array}{ccc|c} c'_1 & 0 & c'_3 & \tilde{c} \\ \hline c_{11} & 0 & c_{13} & 'c_1 \\ c_{21} & c_{22} & c_{23} & 'c_2 \\ c_{31} & 0 & c_{33} & 'c_3 \end{array}\right)$;

3. $\left(\begin{array}{c|c} c' & \tilde{c} \\ \hline c^0 & 'c_1 \end{array}\right)$, $\left(\begin{array}{c|c} c' & \tilde{c} \\ \hline c^0 & 'c \end{array}\right) \cdot \left(\begin{array}{c|c} v^0 & 'v \\ \hline 0 & 1 \end{array}\right)$;

4. $\left(\begin{array}{c|c} 1 & u' \\ \hline 0 & u^0 \end{array}\right) \cdot \left(\begin{array}{c|c} c' & \tilde{c} \\ \hline c^0 & 'c \end{array}\right)$, $\left(\begin{array}{c|c} c' & \tilde{c} \\ \hline c^0 & 'c \end{array}\right)$.

If both matrices from one of these pairs belong to $M(R, \Sigma)$, *then the first matrix in this pair is left adjacent to the second one*.

PROOF. For each of the pairs, the adjacency relation is easy to write. Denote the first matrix in a pair by a, the second by b. In case 1, we should set

$$p = \left(\begin{array}{cc} \hline 0 & 0 \\ 1 & 0 \\ 0 & 0 \\ 0 & 1 \end{array}\right), \quad q = \left(\begin{array}{cc|c} \hline -1 & 0 & 0 \\ 0 & 0 & 0 \\ 0 & -1 & 0 \end{array}\right);$$

in case 2,

$$p = \left(\begin{array}{ccc} \hline 0 & 0 & 0 \\ 1 & 0 & 0 \\ 0 & 0 & 1 \end{array}\right), \quad q = \left(\begin{array}{ccc|c} \hline -1 & 0 & 0 & 0 \\ 0 & 0 & -1 & 0 \end{array}\right),$$

in case 3, $p = \left(\begin{array}{c} \hline 0 \\ 1 \end{array}\right)$, $q = \left(\begin{array}{c|c} \hline -v^0 & -'v \end{array}\right)$, and in case 4, $p = u$, $q = \left(\begin{array}{c|c} \hline -1 & 0 \end{array}\right)$.

The lemma is proved.

Denote by $\sim$ the equivalence relation generated by $\to$. For any pair considered in Lemma 4, substitution of the second matrix in the pair instead of the first one does not change the equivalence class. Such a substitution in case 1 is called cancellation of a trivial row, in case 2 insertion of a trivial column, and in case 3 triangular transformation of columns. Also, substitution of the first matrix instead of the second one is called insertion of a trivial row in case 1, cancellation of a trivial column in case 2, and triangular transformation of rows in case 4.

LEMMA 5. *The relation $\sim$ coincides with the universal Σ-inverting congruence θ_Σ.*

PROOF. $\sim$ is a congruence relation by Lemma 3. It suffices to show that $M/\sim$ is a preadditive category with zero morphisms $(\lfloor {}^m 0^n \rfloor)/\sim$ and identity morphisms $(\lfloor 1_n \rfloor)/\sim$, and that $(\lfloor \underline{a} \rfloor) \odot a^* \sim (\lfloor 1 \rfloor)$, $a^* \odot (\lfloor \underline{a} \rfloor) \sim (\lfloor 1 \rfloor)$ for any matrix $a \in \Sigma$.

Operations $\oplus, \odot$ are associative already on $M(R, \Sigma)$ and have respective unit elements $(\lfloor {}^m 0^n \rfloor)$ and $(\lfloor 1_n \rfloor)$. Relations

$$\begin{pmatrix} a' & 0 \\ a^0 & 0 \end{pmatrix} \cdot (\overline{-1\ 1|0}) + \begin{pmatrix} 0 & 0 \\ 1 & -1 \end{pmatrix} \cdot (a \ominus a) = 0,$$

$$\begin{pmatrix} a' & 0 \\ a^0 & 0 \end{pmatrix} \cdot (\widehat{|0}) + \begin{pmatrix} 0 \\ 1 \end{pmatrix} \cdot (0) = 0$$

prove the equivalence $a \ominus a \sim (\lfloor 0 \rfloor)$. Therefore, the sets ${}^m M^n(R, \Sigma)$ are groups under the operation $\oplus$. Commutativity of these groups follows from the identity $\ominus(a \oplus b) = (\ominus a) \oplus (\ominus b)$. Distributivity follows from the identities

$$((a \odot c) \oplus (b \odot c)) \cdot \hat{q} + \hat{p} \cdot ((a \oplus b) \odot c) = 0,$$
$$(a \odot (b \oplus c)) \cdot \hat{v} + \hat{u} \cdot ((a \odot b) \oplus (a \odot c)) = 0,$$

where

$$p = \begin{pmatrix} \begin{array}{ccc} \overline{0\ 0\ 0} \\ 1\ 0\ 0 \\ 0\ 0\ 1 \\ 0\ 1\ 0 \\ 0\ 0\ 1 \end{array} \end{pmatrix}, \quad q = \begin{pmatrix} \begin{array}{cccc} -1 & 0 & 0 & 0 \\ 0 & 0 & -1 & 0 \\ 0 & -1 & 0 & 0 \\ 0 & 0 & -1 & 0 \end{array} \end{pmatrix},$$

$$u = \begin{pmatrix} \begin{array}{cccc} \overline{0\ 0\ 0\ 0} \\ 1\ 0\ 1\ 0 \\ 0\ 1\ 0\ 0 \\ 0\ 0\ 0\ 1 \end{array} \end{pmatrix}, \quad v = \begin{pmatrix} \begin{array}{cccc} -1 & 0 & -1 & 0 & 0 \\ 0 & -1 & 0 & 0 & 0 \\ 0 & 0 & 0 & -1 & 0 \end{array} \end{pmatrix}.$$

Finally, the identities

$$(1) \cdot (0\ -1) + \widehat{(1|)} \left((\lfloor \underline{a} \rfloor) \odot a^* \right) = 0,$$

$$\left(a^* \odot (\lfloor \underline{a} \rfloor) \right) \cdot \widehat{(|1)} + \begin{pmatrix} 1 \\ 0 \end{pmatrix} \cdot (1) = 0$$

show that the relation $\sim$ is a Σ-inverting congruence. Since $\sim \subseteq \theta_\Sigma$ by the corollary to Lemma 2 and since θ_Σ is the minimal Σ-inverting congruence, it follows that $\sim = \theta_\Sigma$.

The lemma is proved.

The equivalence relation $\sim$ on $M(R, \Sigma)$ can be regarded in an ordinary way as a category $\mathscr{E}$ with elements of $M(R, \Sigma)$ as objects and pairs (a, b) with $a \sim b$ as morphisms.

Due to reflexivity and transitivity, the relation $\to$ defines a subcategory which we denote by $\mathscr{R}$.

We shall introduce three subcategories of $\mathscr{R}$ having the same set of objects as $\mathscr{R}$.

$\mathscr{R}_e$ is defined as a subcategory of pairs of the form

$$\left(\begin{array}{cc|c} a' & a_2 & \tilde{a} \\ \hline a^0 & a_{12} & 'a \\ 0 & a_{22} & 0 \end{array} \right), \quad \left(\begin{array}{c|c} a' & \tilde{a} \\ \hline a^0 & 'a \end{array} \right),$$

where $a^0, a_{22} \in \Sigma$;

$\mathscr{R}_m$ is a subcategory of pairs of the form

$$\left(\begin{array}{c|c} a' & \tilde{a} \\ \hline a^0 & 'a \end{array} \right), \quad \left(\begin{array}{cc|c} 0 & a' & \tilde{a} \\ \hline a_{11} & a_{12} & 'a_1 \\ 0 & a^0 & 'a \end{array} \right),$$

where $a^0, a_{11} \in \Sigma$; $\mathscr{R}_i$ is a subcategory of pairs (a, b) satisfying a relation of the form (4) with matrices p^0, q^0 invertible over R.

Lemmas 3 and 4 imply that the above subsets are indeed subcategories in $\mathscr{R}$ and also that the subcategory $\mathscr{R}_i$ is closed under inversion of morphisms (in the category $\mathscr{E}$).

Let $\mathscr{A}, \mathscr{B}$ be sets of morphisms of the category $\mathscr{E}$, and let, as usual,

$$\mathscr{A} \cdot \mathscr{B} \rightleftharpoons \{\gamma \in \mathscr{E} | (\exists \alpha \in \mathscr{A})(\exists \beta \in \mathscr{B}) \, \gamma = \alpha \cdot \beta\},$$
$$\mathscr{A}^{-1} \rightleftharpoons \{\alpha^{-1} | \alpha \in \mathscr{A}\} = \{(b, a) | (a, b) \in \mathscr{A}\}.$$

LEMMA 6. *We have*:
(a) $\mathscr{R}_m^{-1} \cdot \mathscr{R} \subseteq \mathscr{R} \cdot \mathscr{R}_m^{-1}$,
(b) $\mathscr{R} \cdot \mathscr{R}_e^{-1} \subseteq \mathscr{R}_e^{-1} \cdot \mathscr{R}$,
(c) $\mathscr{R} = \mathscr{R}_m \cdot \mathscr{R}_i \cdot \mathscr{R}_e$,
(d) $\mathscr{E} = \mathscr{R}_e^{-1} \cdot \mathscr{R} \cdot \mathscr{R}_m^{-1}$.

PROOF. (a) A pair (c, b) belongs to the set $\mathscr{R}_m^{-1} \cdot \mathscr{R}$ iff

$$c = \left(\begin{array}{cc|c} 0 & a' & \tilde{a} \\ \hline a_{11} & a_{12} & 'a_1 \\ 0 & a^0 & 'a \end{array} \right), \quad a_{11}, a^0 \in \Sigma,$$

and relation $a\hat{q} + \hat{p}b = 0$ of the form (4) holds. Let

$$p_1 \rightleftharpoons \left(\begin{array}{c|c} 0 & p' \\ \hline 1 & 0 \\ 0 & p^0 \end{array}\right), \qquad q_1 \rightleftharpoons \left(\begin{array}{cc|c} -1 & 0 & 0 \\ 0 & q^0 & 'q \end{array}\right).$$

The row $(b_{11} \ b_{12} \ 'b_1)$ is defined by the equation

$$(a_{11} \ a_{12} \ 'a_1) \cdot \hat{q}_1 + (b_{11} \ b_{12} \ 'b_1) = 0 \tag{5}$$

(this equation defines the size of the unit matrix in q_1). Then for appropriate size of the unit matrix in p_1 we have

$$c \cdot \hat{q}_1 + \hat{p}_1 \cdot \begin{pmatrix} 0 & b' & \tilde{b} \\ b_{11} & b_{12} & 'b_1 \\ 0 & b^0 & 'b \end{pmatrix} = 0,$$

where $b_{11} \in \Sigma$ by (5). Hence, $(c, b) \in \mathscr{R} \cdot \mathscr{R}_m^{-1}$.

Assertion (b) is true due to symmetry between rows and columns in all the definitions. Its proof is similar to the proof of (a), and hence, is omitted.

To prove (c), consider a relation $a\hat{q} + \hat{p}b = 0$ of the form (4). Let

$$c \rightleftharpoons \left(\begin{array}{cc|c} 0 & a' & \tilde{a} \\ b^0 & 0 & 'b \\ 0 & a^0 & 'a \end{array}\right), \qquad d \rightleftharpoons \left(\begin{array}{cc|c} b' & a' & \tilde{b} \\ b^0 & 0 & 'b \\ 0 & a^0 & 0 \end{array}\right),$$

$$p_1 \rightleftharpoons \left(\begin{array}{c|c} p' & 0 \\ \hline 1 & 0 \\ p^0 & 1 \end{array}\right), \qquad q_1 \rightleftharpoons \left(\begin{array}{cc|c} -1 & 0 & 0 \\ q^0 & -1 & 'q \end{array}\right).$$

The equality $c \cdot \hat{q}_1 + \hat{p}_1 \cdot d = 0$ is immediately verified; therefore, $(a, c) \in \mathscr{R}_m$, $(c, d) \in \mathscr{R}_i$, $(d, b) \in \mathscr{R}_e$, and (c) is proved.

To prove (d), it suffices to verify that the set $\mathscr{E}_1 \rightleftharpoons \mathscr{R}_e^{-1} \cdot \mathscr{R} \cdot \mathscr{R}_m^{-1}$ satisfies the conditions $\mathscr{E}_1 \cdot \mathscr{E}_1 \subseteq \mathscr{E}_1$, $\mathscr{E}_1^{-1} \subseteq \mathscr{E}_1$. Using assertions (a)–(c), which have already been verified, we obtain

$$\mathscr{E}_1 \cdot \mathscr{E}_1 = \mathscr{R}_e^{-1} \mathscr{R} \mathscr{R}_m^{-1} \mathscr{R}_e^{-1} \mathscr{R} \mathscr{R}_m^{-1} \subseteq \mathscr{R}_e^{-1} \mathscr{R} \mathscr{R}^{-1} \mathscr{R} \mathscr{R}_m^{-1}$$
$$= \mathscr{R}_e^{-1} \mathscr{R} \mathscr{R}_e^{-1} \mathscr{R}_i \mathscr{R}_m^{-1} \mathscr{R} \mathscr{R}_m^{-1} \subseteq \mathscr{R}_e^{-1} \mathscr{R} \mathscr{R}_m^{-1} = \mathscr{E}_1;$$
$$\mathscr{E}_1^{-1} = \mathscr{R}_m \mathscr{R}_e^{-1} \mathscr{R}_i \mathscr{R}_m^{-1} \mathscr{R}_e \subseteq \mathscr{R} \mathscr{R}_e^{-1} \mathscr{R}_i \mathscr{R}_m^{-1} \mathscr{R} \subseteq \mathscr{R}_e^{-1} \mathscr{R} \mathscr{R}_m^{-1} = \mathscr{E}_1.$$

The lemma is proved.

COROLLARY. *Matrices $a, b \in M(R, \Sigma)$ are equivalent if and only if there exists a relation of the form*

$$\begin{pmatrix} a' & * & * & * \\ a^0 & * & * & * \\ 0 & \Sigma & * & * \end{pmatrix} \cdot \begin{pmatrix} * & * & * \\ * & * & * \\ \Sigma & * & * \\ 0 & b^0 & b' \end{pmatrix} = \begin{pmatrix} 0 & -b' & \tilde{a} - \tilde{b} \\ 0 & 0 & 'a \\ 0 & 0 & 0 \end{pmatrix},$$

where Σ denotes some matrices from Σ and $*$ some matrices of appropriate size.

PROOF. Using Lemma 2 we see that the above assertion is a reformulation of (d) in Lemma 6.

Now we state the main theorem on the construction of localization.

THEOREM 4. *Let R be a ring, let Σ be a multiplicative set of matrices over R. Let*

$$M \rightleftharpoons M(R, \Sigma) \rightleftharpoons \left\{ a = \left(\begin{array}{c|c} a' & \tilde{a} \\ \hline a^0 & 'a \end{array} \right) \middle| a^0 \in \Sigma, \ a', \tilde{a}, 'a \in \mathrm{Mat}(R) \right\},$$

$${}^m M^n \rightleftharpoons {}^m M^n(R, \Sigma) \rightleftharpoons \{ a \in M \mid \tilde{a} \in {}^m R^n \}, \qquad m, n \in \mathbb{N}.$$

Define on the set $M(R, \Sigma)$ the partial operations

$$\begin{cases} {}^m M^n \times {}^m M^n \to {}^m M^n, \qquad m, n \in \mathbb{N}, \\ (a, b) \to a \oplus b = \left(\begin{array}{cc|c} a' & b' & \tilde{a} + \tilde{b} \\ \hline a^0 & 0 & 'a \\ 0 & b^0 & 'b \end{array} \right), \\ {}^m M^n \times {}^n M^k \to {}^m M^k, \qquad m, n, k \in \mathbb{N}, \\ (a, b) \to a \odot b = \left(\begin{array}{cc|c} a' & \tilde{a}b' & \widetilde{ab} \\ \hline a^0 & 'ab' & '\widetilde{ab} \\ 0 & b^0 & 'b \end{array} \right), \end{cases}$$

the mapping

$$\varepsilon : \begin{cases} \mathrm{Mat}(R) \to M(R, \Sigma), \\ {}^m R^n \to {}^m M^n(R, \Sigma), \\ a \to (\underline{|a}), \end{cases}$$

and the relation

$$\left(\begin{array}{c|c} a' & \tilde{a} \\ \hline a^0 & 'a \end{array} \right) \sim \left(\begin{array}{c|c} b' & \tilde{b} \\ \hline b^0 & 'b \end{array} \right)$$

$$\iff \left(\begin{array}{cccc} a' & * & * & * \\ a^0 & * & * & * \\ 0 & \Sigma & * & * \end{array} \right) \cdot \left(\begin{array}{ccc} * & * & * \\ * & * & * \\ \Sigma & * & * \\ 0 & b^0 & 'b \end{array} \right) = \left(\begin{array}{ccc} 0 & -b' & \tilde{a} - \tilde{b} \\ 0 & 0 & 'a \\ 0 & 0 & 0 \end{array} \right).$$

Then $\sim$ is a congruence under the operations $\oplus$, $\odot$, and there exists a one-to-one correspondence

$$\left(\begin{array}{c|c} a' & \tilde{a} \\ \hline a^0 & 'a \end{array} \right) \Big/ \sim \; \longleftrightarrow \; [a] = \left[\begin{array}{c|c} a' & \tilde{a} \\ \hline a^0 & 'a \end{array} \right]$$

between congruence classes and matrices over a ring $R_1 = R\Sigma^{-1}$. Moreover, elements of sets ${}^mM^n(R, \Sigma)$ correspond to matrices with m rows and n columns, elements $a \oplus b$, $c \odot d$ correspond to matrices $[a]+[b]$, $[c]\cdot[d]$, the mapping

$$u_\Sigma = {}^1\varepsilon^1 / \sim : \begin{cases} R \to R_1 \\ r \to (\underline{r}) \end{cases}$$

is the universal Σ-inverting homomorphism, and the element

$$a^* = \left(\begin{array}{c|c} -1 & 0 \\ \hline a & 1 \end{array}\right), \qquad a \in \Sigma,$$

corresponds to the matrix $u_\Sigma(a)^{-1}$.

COROLLARY 1. *The following facts about a matrix $a \in M(R, \Sigma)$ are equivalent under the assumptions of Theorem 4:*

(a) *a represents a zero matrix over $R\Sigma^{-1}$.*

(b) *There exists a relation of the form*

$$\begin{pmatrix} a' & * & * \\ a^0 & * & * \\ 0 & \Sigma & * \end{pmatrix} \cdot \begin{pmatrix} * & * \\ * & * \\ \Sigma & * \end{pmatrix} = \begin{pmatrix} 0 & \tilde{a} \\ 0 & {}'a \\ 0 & 0 \end{pmatrix}.$$

(c) *There exists a relation of the form*

$$\begin{pmatrix} * & * & * \\ \Sigma & * & * \end{pmatrix} \cdot \begin{pmatrix} * & * & * \\ \Sigma & * & * \\ 0 & a^0 & {}'a \end{pmatrix} = \begin{pmatrix} 0 & a' & \tilde{a} \\ 0 & 0 & 0 \end{pmatrix}.$$

COROLLARY 2. *Under the assumptions of Theorem 4, the kernel of the universal Σ-inverting homomorphism $u_\Sigma : R \to R\Sigma^{-1}$ consists precisely of the elements $r \in R$ satisfying a relation of the form*

$$\begin{pmatrix} a' & \tilde{a} \\ a^0 & {}'a \end{pmatrix} \cdot \begin{pmatrix} b' & \tilde{b} \\ b^0 & {}'b \end{pmatrix} = \begin{pmatrix} 0 & r \\ 0 & 0 \end{pmatrix}, \qquad a^0, b^0 \in \Sigma.$$

Corollary 2 implies the following result.

THEOREM 5. *A multiplicative set Σ is potentially invertible if and only if all the quasi-identities of the form*

$$\begin{pmatrix} a' & \tilde{a} \\ a^0 & {}'a \end{pmatrix} \cdot \begin{pmatrix} b' & \tilde{b} \\ b^0 & {}'b \end{pmatrix} = \begin{pmatrix} 0 & c \\ 0 & 0 \end{pmatrix} \Longrightarrow c = 0$$

hold over R. Here $a^0, b^0 \in \Sigma$, $c \in \Sigma$ and the remaining matrices have appropriate sizes.

Let Σ be a multiplicative set of matrices over a ring R. We are going to describe zero divisors and invertible elements in the category $\mathrm{Mat}(R, \Sigma^{-1})$.

LEMMA 7. *Let* $a, b, c \in M(R, \Sigma)$, $(c, a \odot b) \in \mathscr{R}_e$. *Then there exist matrices* $a_1, b_1 \in M(R, \Sigma)$ *such that*
$$(a_1, a) \in \mathscr{R}_e, \qquad (b_1, b) \in \mathscr{R}_e, \qquad a_1 \odot b_1 \to c.$$

PROOF. We have
$$c = \left(\begin{array}{ccc|c} a' & \tilde{a}b' & c_1 & \tilde{a}\tilde{b} \\ \hline a^0 & 'ab' & c_2 & 'a\tilde{b} \\ 0 & b^0 & c_3 & 'b \\ 0 & 0 & c^0 & 0 \end{array} \right), \qquad c^0 \in \Sigma.$$

Let
$$a_1 \rightleftharpoons \left(\begin{array}{cc|c} a' & c_1 & \tilde{a} \\ \hline a^0 & c_2 & 'a \\ 0 & c^0 & 0 \end{array} \right), \qquad b_1 \rightleftharpoons \left(\begin{array}{cc|c} b' & 0 & \tilde{b} \\ \hline b^0 & c_3 & 'b \\ 0 & c^0 & 0 \end{array} \right),$$

$$p \rightleftharpoons \left(\begin{array}{ccc} 0 & 0 & 0 \\ \hline 1 & 0 & 0 \\ 0 & 0 & 1 \\ 0 & 1 & 0 \\ 0 & 0 & 1 \end{array} \right), \qquad q \rightleftharpoons \left(\begin{array}{ccc|c} -1 & 0 & 0 & 0 \\ 0 & 0 & -1 & 0 \\ 0 & -1 & 0 & 0 \\ 0 & 0 & -1 & 0 \end{array} \right).$$

If the sizes of unit matrices in p and q are properly chosen, the equality $(a_1 \odot b_1)\hat{q} + \hat{p}c = 0$ holds.

The lemma is proved.

The following dual statement is proved similarly.

LEMMA 7'. *Let* $a, b, c \in M(R, \Sigma)$, $(a \odot b, c) \in \mathscr{R}_m$. *Then there exist matrices* $a_1, b_1 \in M(R, \Sigma)$ *such that*
$$(a, a_1) \in \mathscr{R}_m, \qquad (b, b_1) \in \mathscr{R}_m, \qquad c \to a_1 \odot b_1.$$

PROPOSITION 3. *Let* Σ *be a multiplicative set of matrices over a ring* R. *Then for any relation*
$$u \cdot v = [\underline{\mid r}], \qquad u, v \in \mathrm{Mat}(R, \Sigma^{-1}), \qquad r \in \mathrm{Mat}(R)$$
there exist matrices $a, b \in \mathrm{Mat}(R, \Sigma)$ *satisfying*
$$\left(\begin{array}{ccc} a' & \tilde{a} & * \\ a^0 & 'a & * \end{array} \right) \cdot \left(\begin{array}{cc} * & * \\ b' & \tilde{b} \\ b^0 & 'b \end{array} \right) = \left(\begin{array}{cc} 0 & r \\ 0 & 0 \end{array} \right) \qquad (6)$$
such that $[a] = u$, $[b] = v$.

PROOF. Let $u = [a_2]$, $-v = [b_2]$, $a_2, b_2 \in M(R, \Sigma)$. By Lemma 6 there exist matrices $c, d \in M(R, \Sigma)$ such that
$$(c, a_2 \odot b_2) \in \mathscr{R}_e, \qquad c \to d, \qquad ((\underline{\mid -r}), d) \in \mathscr{R}_m.$$

By Lemma 7, there exist matrices $a_1, b_1 \in M(R, \Sigma)$ such that
$$(a_1, a_2) \in \mathscr{R}_e, \quad (b_1, b_2) \in \mathscr{R}_e, \quad a_1 \odot b_1 \to c.$$

By Lemma 3, $a_1 \odot b_1 \to d$, where d has the form $\left(\begin{array}{c|c} 0 & -r \\ \hline d^0 & 'd \end{array}\right)$. Consider the relation between matrices $a_1 \odot b_1$ and d:

$$\begin{pmatrix} a_1' & \tilde{a}_1 b_1' & \tilde{a}_1 \tilde{b}_1 \\ a_1^0 & 'a_1 b_1' & 'a_1 \tilde{b}_1 \\ 0 & b_1^0 & 'b_1 \end{pmatrix} \cdot \begin{pmatrix} q_{11} & q_{12} \\ q_{21} & q_{22} \\ 0 & -1 \end{pmatrix} + \begin{pmatrix} 1 & p_1 \\ 0 & p_2 \\ 0 & p_3 \end{pmatrix} \cdot \begin{pmatrix} 0 & -r \\ d^0 & 'd \end{pmatrix} = 0,$$

which immediately implies the following relations:

$$\begin{pmatrix} a_1' & \tilde{a}_1 & p_1 \\ a_1^0 & 'a_1 & p_2 \end{pmatrix} \cdot \begin{pmatrix} q_{11} & q_{12} \\ b_1' q_{21} & b_1' q_{22} - \tilde{b} \\ d^0 & 'd \end{pmatrix} = \begin{pmatrix} 0 & r \\ 0 & 0 \end{pmatrix}, \qquad (7)$$

$$\begin{pmatrix} b_1' & \tilde{b}_1 \\ b_1^0 & 'b_1 \end{pmatrix} \cdot \begin{pmatrix} q_{21} & q_{22} \\ 0 & -1 \end{pmatrix} + \begin{pmatrix} 1 & 0 \\ 0 & p_3 \end{pmatrix} \cdot \begin{pmatrix} -b_1' q_{21} & -b_1' q_{22} + \tilde{b} \\ d^0 & 'd \end{pmatrix} = 0. \quad (8)$$

Set
$$a \rightleftharpoons a_1, \qquad b \rightleftharpoons \left(\begin{array}{c|c} b_1' q_{21} & b_1' q_{22} - \tilde{b} \\ \hline d^0 & 'd \end{array}\right).$$

By (8) we have $b \sim (\ominus b_1) \sim (\ominus b_2)$. Hence, $[a] = u$, $[b] = v$, and the relation (7) has the required form (6).

The proposition is proved.

§2. Localizations by independent sets of matrices

Let Σ be a set (not necessarily multiplicative) of matrices over a fixed ring R. We call any relation $a \cdot b = 0$ of the form

$$\left(\begin{array}{c|c|c} a' & x_1 & u_1 \\ \hline a^0 & x_2 & u_2 \end{array}\right) \cdot \left(\begin{array}{c|c} v_1 & v_2 \\ \hline y_1 & y_2 \\ \hline b^0 & 'b \end{array}\right) = 0 \qquad (1)$$

a Σ-*dependence* relation.

In more rigorous terms, a Σ-dependence relation is given if there exist 12 matrices (participating in (1)) satisfying the four relations:

$$a' \cdot v_1 + x_1 \cdot y_1 + u_1 \cdot b^0 = 0, \qquad a' \cdot v_2 + x_1 \cdot y_2 + u_1 \cdot 'b = 0,$$
$$a^0 \cdot v_1 + x_2 \cdot y_1 + u_2 \cdot b^0 = 0, \qquad a^0 \cdot v_2 + x_2 \cdot y_2 + u_2 \cdot 'b = 0,$$

where matrix sizes allow us to assemble matrices a and b of the form shown in (1) such that $a \cdot b = 0$.

Some of the 12 matrices in (1) might be empty. In this case we shall use shorthand notation similar to that of § 1.

The number of columns in the matrix $x = \begin{pmatrix} x_1 \\ x_2 \end{pmatrix}$ is equal to the number of rows in the matrix $y = (y_1 \ y_2)$ and is called the length of relation (1).

If (1) is of the form

$$\left(\begin{array}{c|cc|c} a' & x_{11} & 0 & 0 \\ \hline a^0 & x_{21} & 0 & 0 \end{array} \right) \cdot \left(\begin{array}{c|c} 0 & 0 \\ \hline 0 & 0 \\ \hline y_{21} & y_{22} \\ \hline b^0 & {}'b \end{array} \right) = 0,$$

then it is called *trivial*.

Let, in addition to (1), there exist a pair of mutually inverse matrices

$$\alpha = \left(\begin{array}{c|ccc|c} \alpha_{11} & \alpha_{12} & \alpha_{13} & \alpha_{14} \\ \hline \alpha_{21} & \alpha_{22} & \alpha_{23} & \alpha_{24} \\ \hline \alpha_{31} & \alpha_{32} & \alpha_{33} & \alpha_{34} \end{array} \right), \quad \beta = \left(\begin{array}{c|cc|c} \beta_{11} & \beta_{12} & \beta_{13} \\ \hline \beta_{21} & \beta_{22} & \beta_{23} \\ \beta_{31} & \beta_{32} & \beta_{33} \\ \hline \beta_{41} & \beta_{42} & \beta_{43} \end{array} \right) \quad (2)$$

such that $a\alpha \cdot \beta b = 0$ is a trivial Σ-dependence relation of the same length as the original relation $a \cdot b = 0$. In this case we say that the pair (α, β) trivializes (or, more precisely, Σ-trivializes) the relation (1).

Matrices α, β are partitioned in such a way that multiplication is performed blockwise and the partitions agree with each other.

Denote by Σ_0 the set $\Sigma \cup \{1_0\}$ obtained by adjoining the empty (0×0) matrix. A set Σ is called *n-independent* if any Σ_0-dependence relation of length $\leq n$ can be Σ_0-trivialized. 0-independent sets will be called simply *independent*.

Note that independent sets do not contain zero-divisors (since relations of the form $(\overline{a^0 \|}) \cdot (\underline{\| v}) = 0$, $(\underline{\| u}) \cdot (\overline{\| b^0}) = 0$ are trivializable).

LEMMA 8. *Let Σ be an independent multiplicative set of matrices over R, $a \in M(R, \Sigma)$. The following statements are equivalent:*

(a) *matrix a represents the zero matrix over $R\Sigma^{-1}$;*

(b) *there exists a relation of the form*

$$\begin{pmatrix} a' & \tilde{a} \\ a^0 & {}'a \end{pmatrix} = \begin{pmatrix} b' \\ b^0 \end{pmatrix} \cdot (c^0 \ {}'c), \quad (3)$$

where $b^0 \in \Sigma$;

(c) *there exists a relation of the form (3), where $c^0 \in \Sigma$.*

PROOF. Implications $(b) \Rightarrow (a)$, $(c) \Rightarrow (a)$ follow from Lemma 1. Let us prove that (a) implies (b). By Corollary 1 to Theorem 2, there exists a

relation of the form
$$\begin{pmatrix} a' & p_1 & a_{12} \\ a^0 & p_2 & a_{22} \\ 0 & p_3 & a_{32} \end{pmatrix} \cdot \begin{pmatrix} b_{11} & b_{12} \\ q_1 & q_2 \\ b_{21} & b_{22} \end{pmatrix} = \begin{pmatrix} 0 & \tilde{a} \\ 0 & {'a} \\ 0 & 0 \end{pmatrix}, \quad p_3, b_{21} \in \Sigma.$$

The Σ-dependence relation $(\overline{p_3 \| a_{32}}) \cdot \left(\dfrac{q_1}{b_{21}}\Big\|\right) = 0$ is known to be trivializable; we can assume it is trivial, i.e., $a_{32} = 0$, $q_1 = 0$. Then $q_2 = 0$, since p_3 is not a zero-divisor. Therefore,
$$\begin{pmatrix} a' & a_{12} \\ a^0 & a_{22} \end{pmatrix} \cdot \begin{pmatrix} b_{11} & b_{12} \\ b_{21} & b_{22} \end{pmatrix} = \begin{pmatrix} 0 & \tilde{a} \\ 0 & {'a} \end{pmatrix}.$$

Let (α, β) be a pair of matrices trivializing the relation
$$\left(\begin{array}{c|c} a' & a_{12} \\ a^0 & a_{22} \end{array}\right) \cdot \left(\dfrac{b_{11}}{b_{21}}\Big\|\right) = 0,$$

and let
$$\begin{pmatrix} a' & a_{12} \\ a^0 & a_{22} \end{pmatrix} \cdot \begin{pmatrix} \alpha_{11} & \alpha_{12} \\ \alpha_{21} & \alpha_{22} \end{pmatrix} = \begin{pmatrix} b' & 0 \\ b^0 & 0 \end{pmatrix},$$
$$\begin{pmatrix} \beta_{11} & \beta_{12} \\ \beta_{21} & \beta_{22} \end{pmatrix} \cdot \begin{pmatrix} b_{11} & b_{12} \\ b_{21} & b_{22} \end{pmatrix} = \begin{pmatrix} 0 & \tilde{d} \\ d^0 & {'d} \end{pmatrix}.$$

Then
$$\begin{pmatrix} a' & \tilde{a} \\ a^0 & {'a} \end{pmatrix} = \begin{pmatrix} b' \\ b^0 \end{pmatrix} \cdot (\beta_{11} \quad \tilde{d}),$$

which is the desired assertion.

$(a) \Rightarrow (b)$ can be proved similarly.

The lemma is proved.

COROLLARY. *Any independent multiplicative set Σ of matrices over an arbitrary ring R is potentially invertible.*

PROOF. Let $r \in R$ be an element of the kernel of the universal Σ-inverting homomorphism. By Corollary 2 of Theorem 4, there exists a relation of the form
$$\begin{pmatrix} a' & \tilde{a} \\ a^0 & {'a} \end{pmatrix} \cdot \begin{pmatrix} b' & \tilde{b} \\ b^0 & {'b} \end{pmatrix} = \begin{pmatrix} 0 & r \\ 0 & 0 \end{pmatrix}, \quad a^0, b^0 \in \Sigma.$$

We can assume ${'a} = 0$, $b' = 0$. In this case $\tilde{a} = 0$, $\tilde{b} = 0$, since a^0, b^0 are not zero divisors. This implies $r = 0$.

The assertion is proved.

Note that the above assertion can be formally deduced from Lemma 8.

If there exists an n-independent set of matrices over R, then all relations of length $\leq n$ of the form
$$(\lfloor x \rfloor) \cdot (\overline{\lfloor y \rfloor}) = 0$$
are trivializable, and therefore, R is an n-FI-ring. (All definitions and results about n-FI-rings used here can be found in [9].)

PROPOSITION 4. *Let Σ be an n-independent multiplicative set of matrices over R. Then $R\Sigma^{-1}$ is an n-FI-ring.*

PROOF. Let u be a row of length n, v be a column of length n over the ring $R\Sigma^{-1}$, and let $u \cdot v = 0$. It suffices to prove that there exists a pair of mutually inverse square matrices

$$[\gamma], [\delta] \in \text{Mat}(R\Sigma^{-1}), \quad \gamma, \delta \in M(R, \Sigma),$$

such that either the last element of the row $u \cdot [\gamma]$ equals zero or the last element of the column $[\delta] \cdot v$ equals zero.

By Proposition 3 there exist matrices $a \in {}^1M^n(R, \Sigma)$, $b \in {}^nM^1(R, \Sigma)$ satisfying some relation of the form

$$\left(\begin{array}{c|c|c} a' & \tilde{a} & p_1 \\ \hline a^0 & 'a & p_2 \end{array} \right) \cdot \left(\begin{array}{c|c} q_1 & q_2 \\ \hline b' & \tilde{b} \\ \hline b^0 & 'b \end{array} \right) = 0$$

and such that $[a] = u$, $[b] = v$. Let (α, β) be a pair of mutually inverse matrices of the form (2) trivializing the last relation. Then for some matrices $c, d \in M(R, \Sigma)$ we have

$$\left(\begin{array}{c|c|c} a' & \tilde{a} & p_1 \\ \hline a^0 & 'a & p_2 \end{array} \right) \alpha = \left(\begin{array}{c|c|c|c} c' & \tilde{c} & 0 & 0 \\ \hline c^0 & 'c & 0 & 0 \end{array} \right), \quad \beta \left(\begin{array}{c|c} q_1 & q_2 \\ \hline b' & \tilde{b} \\ \hline b^0 & 'b \end{array} \right) = \left(\begin{array}{c|c} 0 & 0 \\ \hline 0 & 0 \\ \hline d' & \tilde{d} \\ \hline d^0 & 'd \end{array} \right).$$

Let

$$\gamma \rightleftharpoons \left(\begin{array}{cc|cc} b' & \alpha_{21} & \alpha_{22} & \alpha_{23} \\ b^0 & \alpha_{31} & \alpha_{32} & \alpha_{33} \\ 0 & c^0 & 'c & 0 \end{array} \right), \quad \delta \rightleftharpoons \left(\begin{array}{cc|cc} 0 & \beta_{21} & \beta_{22} \\ d' & \beta_{31} & \beta_{32} \\ d^0 & \beta_{41} & \beta_{42} \\ 0 & a^0 & 'a \end{array} \right).$$

We want to verify that matrices $[\gamma]$ and $[\delta]$ are mutually inverse. We have

$$\delta \odot \gamma = \left(\begin{array}{cccc|cc} 0 & \beta_{21} & \beta_{22}b' & \beta_{22}\alpha_{21} & \beta_{22}\alpha_{22} & \beta_{22}\alpha_{23} \\ d' & \beta_{31} & \beta_{32}b' & \beta_{32}\alpha_{21} & \beta_{32}\alpha_{22} & \beta_{32}\alpha_{23} \\ d^0 & \beta_{41} & \beta_{42}b' & \beta_{42}\alpha_{21} & \beta_{42}\alpha_{22} & \beta_{42}\alpha_{23} \\ 0 & a^0 & 'ab' & 'a\alpha_{21} & 'a\alpha_{22} & 'a\alpha_{23} \\ 0 & 0 & b^0 & \alpha_{31} & \alpha_{32} & \alpha_{33} \\ 0 & 0 & 0 & c^0 & 'c & 0 \end{array} \right).$$

Apply Lemma 4. After multiplying the resulting matrix on the left and on the right by respective matrices

$$\begin{pmatrix} 1 & 0 & 0 & 0 & \beta_{23} & 0 \\ 0 & 1 & 0 & 0 & \beta_{33} & 0 \\ 0 & 0 & 1 & 0 & \beta_{43} & 0 \\ 0 & 0 & 0 & 1 & p_1 & -1 \\ 0 & 0 & 0 & 0 & 1 & 0 \\ 0 & 0 & 0 & 0 & 0 & 1 \end{pmatrix}, \quad \begin{pmatrix} 1 & 0 & -1 & 0 & 0 & 0 \\ 0 & 1 & q_1 & \alpha_{11} & \alpha_{12} & \alpha_{13} \\ 0 & 0 & 1 & 0 & 0 & 0 \\ 0 & 0 & 0 & 1 & 0 & 0 \\ 0 & 0 & 0 & 0 & 1 & 0 \\ 0 & 0 & 0 & 0 & 0 & 1 \end{pmatrix},$$

we obtain

$$\left(\begin{array}{cccc|cc} 0 & \beta_{21} & 0 & 0 & 1 & 0 \\ d' & \beta_{31} & 0 & 0 & 0 & 1 \\ \hline d^0 & \beta_{41} & 0 & 0 & 0 & 0 \\ 0 & a^0 & 0 & 0 & 0 & 0 \\ 0 & 0 & b^0 & \alpha_{31} & \alpha_{32} & \alpha_{33} \\ 0 & 0 & 0 & c^0 & {'c} & 0 \end{array} \right),$$

and after cancellation of trivial rows and columns this matrix can be reduced to

$$\left(\begin{array}{c|cc} & 1 & 0 \\ & 0 & 1 \end{array} \right).$$

Similarly, the matrix $\gamma \odot \delta$ can be reduced to the unit matrix by triangular transformations and cancellations. Thus, $\delta \odot \gamma \sim (\underline{\;1\;})$, $\gamma \odot \delta \sim (\underline{\;1\;})$.

Next we shall verify that

$$a \odot \gamma \sim \left(\begin{array}{c|cc} c' & \tilde{c} & 0 \\ \hline c^0 & {'c} & 0 \end{array} \right), \quad \delta \odot b \sim \left(\begin{array}{c|c} 0 & 0 \\ \hline d' & \tilde{d} \\ d^0 & {'d} \end{array} \right). \tag{4}$$

We have

$$\left(\begin{array}{cc|ccc} a' & \tilde{a}b' & \tilde{a}\alpha_{21} & \tilde{a}\alpha_{22} & \tilde{a}\alpha_{23} \\ \hline a^0 & {'a}b' & {'a}\alpha_{21} & {'a}\alpha_{22} & {'a}\alpha_{23} \\ 0 & b^0 & \alpha_{31} & \alpha_{32} & \alpha_{33} \\ 0 & 0 & c^0 & {'c} & 0 \end{array} \right).$$

Multiplying the resulting matrix on the left and on the right, respectively, by

$$\begin{pmatrix} 1 & 0 & p_1 & 0 \\ 0 & 1 & p_2 & -1 \\ 0 & 0 & 1 & 0 \\ 0 & 0 & 0 & 1 \end{pmatrix}, \quad \begin{pmatrix} 1 & q_1 & \alpha_{11} & \alpha_{12} & \alpha_{13} \\ 0 & 1 & 0 & 0 & 0 \\ 0 & 0 & 1 & 0 & 0 \\ 0 & 0 & 0 & 1 & 0 \\ 0 & 0 & 0 & 0 & 1 \end{pmatrix}$$

we get
$$\left(\begin{array}{cc|cc|c} a' & 0 & c' & \tilde{c} & 0 \\ \hline a^0 & 0 & 0 & 0 & 0 \\ 0 & b^0 & \alpha_{31} & \alpha_{32} & \alpha_{33} \\ 0 & 0 & c^0 & 'c & 0 \end{array}\right)$$
which after cancellations is reduced to
$$\left(\begin{array}{c|cc} c' & \tilde{c} & 0 \\ \hline c^0 & 'c & 0 \end{array}\right).$$

The second equivalence in (4) is proved similarly. The proposition is proved.

LEMMA 9. *The multiplicative closure $\bar{\Sigma}$ of an n-independent set Σ is n-independent.*

PROOF. We can assume that the empty matrix 1_0 belongs to Σ. Consider a $\bar{\Sigma}$-dependence relation $a \cdot b = 0$ of the form (1) and prove that it is trivializable. We shall argue by induction on the sum of the number of rows in a^0 and the number of columns in b^0. If both matrices a^0, b^0 belong to Σ, then the relation is trivializable by the assumption of the lemma. Assume one matrix, say a^0, does not belong to Σ. Then (1) can be rewritten in the form

$$\left(\begin{array}{cc|c|c} a_{11} & a_{12} & x_1 & u_1 \\ \hline a_{21} & a_{22} & x_2 & u_2 \\ 0 & a_{32} & x_3 & u_3 \end{array}\right) \cdot \left(\begin{array}{cc|c|c} v_{11} & v_{21} & y_1 & b^0 \\ \hline v_{12} & v_{22} & y_2 & 'b \end{array}\right)^T = 0, \qquad a_{21}, a_{32} \in \Sigma$$

(from here on the superscript T means formal transposition, i.e., transposition that is not applied to blocks; e.g., $\begin{pmatrix} a & b \\ c & d \end{pmatrix}^T = \begin{pmatrix} a & c \\ b & d \end{pmatrix}$, where a, b, c, d are arbitrary matrices). The relation

$$\left(\begin{array}{c|c|c} \overline{a_{32}} & x_3 & u_3 \end{array}\right) \cdot \left(\begin{array}{c|c|c} v_{21} & y_1 & b^0 \\ \hline v_{22} & y_2 & 'b \end{array}\right)^T = 0$$

is trivializable by the induction assumption, and we can assume it to be trivial. Therefore, the relation in question has the form

$$\left(\begin{array}{cc|cc|c} a_{11} & a_{12} & x_{11} & x_{12} & u_1 \\ \hline a_{21} & a_{22} & x_{21} & x_{22} & u_2 \\ 0 & a_{32} & x_{31} & 0 & 0 \end{array}\right) \cdot \left(\begin{array}{cc|cc|c} v_{11} & 0 & 0 & y_{21} & b^0 \\ \hline v_{12} & 0 & 0 & y_{22} & 'b \end{array}\right)^T = 0, \qquad (5)$$

and this yields the relation

$$\left(\begin{array}{c|c|c} a_{11} & x_{12} & u_1 \\ \hline a_{21} & x_{22} & u_2 \end{array}\right) \cdot \left(\begin{array}{c|c|c} v_{11} & y_{21} & b^0 \\ \hline v_{12} & y_{22} & 'b \end{array}\right)^T = 0,$$

which is trivializable by some pair (α, β) due to the induction assumption. By inserting rows and columns of unit matrices into appropriate places, we can obtain a pair which trivializes (5).

The lemma is proved.

LEMMA 10. *A set Σ of matrices over an n-FI-ring R is n-independent if and only if all relations*

$$(\overline{a|x}) \cdot \left(\overline{\dfrac{v}{y}}\middle|\right) = 0, \quad a \in \Sigma;$$

$$(\underline{|x|u}) \cdot \left(\middle|\overline{\dfrac{y}{b}}\right) = 0, \quad b \in \Sigma;$$

$$(\overline{a|x|u}) \cdot (v|y|b)^{\mathrm{T}} = 0, \quad a, b \in \Sigma$$

with length not exceeding n are trivializable.

PROOF. Consider a Σ_0-dependence relation $a \cdot b = 0$ of the form (1) with length $\leq n$. We can assume that the relation

$$\left(\overline{a^0 | x_2 | u_2}\right) \cdot \left(v_1 | y_1 | b^0\right)^{\mathrm{T}} = 0$$

is trivial, and hence, that the relation in question has the form

$$\left(\begin{array}{c|ccc|c} a' & x_{11} & x_{12} & u_1 \\ \hline a^0 & x_{21} & 0 & 0 \end{array}\right) \cdot \left(\begin{array}{c|ccc|c} 0 & 0 & y_{21} & b^0 \\ \hline v_2 & y_{12} & y_{22} & 'b \end{array}\right)^{\mathrm{T}} = 0.$$

In this case the relations

$$(\underline{|x_{12}|u_1}) \cdot (\underline{|y_{21}|b^0})^{\mathrm{T}} = 0, \quad (\overline{a^0|x_{21}|}) \cdot (\overline{v_2|y_{12}|})^{\mathrm{T}} = 0$$

hold and can also be assumed trivial. Consequently, the relation in question has the form

$$\left(\begin{array}{c|ccccc} a' & x_{11} & x_{12} & x_{13} & 0 & 0 \\ \hline a^0 & x_{12} & 0 & 0 & 0 & 0 \end{array}\right) \cdot \left(\begin{array}{cc|cccc} 0 & 0 & 0 & 0 & y_{41} & b^0 \\ \hline 0 & 0 & y_{22} & y_{32} & y_{42} & 'b \end{array}\right)^{\mathrm{T}} = 0,$$

and we arrive at the relation

$$(\underline{|x_{13}|}) \cdot (\overline{|y_{32}}) = 0,$$

which is trivializable by the definition of an n-FI-ring.

The lemma is proved.

COROLLARY. *A set Σ of matrices over R is independent if and only if it has no zero divisors and all relations of the form*

$$(\overline{a \| u}) \cdot \left(\dfrac{v}{b}\middle\|\right) = 0, \quad a, b \in \Sigma,$$

are Σ-trivializable.

Lemmas 9 and 10 allow us to drop the multiplicativity requirements in the corollary to Lemma 8 and in Proposition 4.

Our results can be stated as follows.

THEOREM 6. *Let Σ be a set of rectangular matrices over a ring R. Let it contain no zero-divisors and satisfy the following property*:
For any relation
$$av + ub = 0, \quad a, b \in \Sigma,$$
there exists a pair of mutually inverse matrices over R
$$\alpha = \begin{pmatrix} \alpha_{11} & \alpha_{12} \\ \alpha_{21} & \alpha_{22} \end{pmatrix}, \quad \beta = \begin{pmatrix} \beta_{11} & \beta_{12} \\ \beta_{21} & \beta_{22} \end{pmatrix}$$
such that
$$a\alpha_{11} + u\alpha_{21} \in \Sigma, \quad a\alpha_{12} + u\alpha_{22} = 0,$$
$$\beta_{11}v + \beta_{12}b = 0, \quad \beta_{21}v + \beta_{22}b \in \Sigma.$$
Then Σ is a potentially invertible set.

Next, let Σ be the set of all nonzero elements (i.e., of (1×1) matrices) over a 2-FI-ring R. By the definition of a 2-FI-ring, the assumptions of Theorem 6 are satisfied, so that the following statement holds:

THEOREM 7. *Any 2-FI-ring is invertible.*

THEOREM 8. *Let $\Sigma \subseteq \mathrm{Mat}(R)$ be a set of rectangular matrices over R such that any relation*
$$a \cdot v + x \cdot y + u \cdot b = 0, \quad x \in R^m, \ y \in {}^m R, \ m \leq n, \ a, b \in \Sigma \cup \{1_0\} \rightleftharpoons \Sigma_0,$$
is Σ_0-trivializable. Then $R\Sigma^{-1}$ is an n-FI-ring.

Denote by $\Sigma_l(R)$ the set of all full square matrices of order $l > 0$ over a k-FI-ring R. (A matrix a of order l is called full if it cannot be represented in the form $a = bc$, $b \in {}^l R^{l_1}$, $c \in {}^{l_1} R^l$, $l_1 < l$.) If $k \geq 2l$, then the set $\Sigma_l(R)$ satisfies the conditions of Theorem 6 and the conditions of Theorem 8 for $n = k - 2l$, which immediately follow from the definition of the k-FI-ring (see [9, §1.1]). This yields the following theorem.

THEOREM 9. *Let R be a k-FI-ring. If $l \leq k/2$, then the set $\Sigma_l(R)$ is potentially invertible and the universal $\Sigma_l(R)$-inverting ring $R\Sigma_l(R)^{-1}$ is a $(k - 2l)$-FI-ring.*

This theorem provides an answer to the question of G. Bergman [22, p. 77]. For $l = 1$, $\Sigma_l(R)$ coincides with the set R^* of all nonzero elements of R. This yields the following result.

COROLLARY. *Let R be a k-FI-ring, $k \geq 3$. Then $R(R^*)^{-1}$ is a $(k-2)$-FI-ring.*

References

1. S. I. Adyan, *Defining relations and algorithmic problems for groups and semigroups*, Trudy Mat. Inst. Steklov. **85** (1966); English transl., Proc. Steklov Inst. Math. **1967**.
2. L. A. Bokut', *The imbeddings of rings in skew fields*, Dokl. Akad. Nauk SSSR **175** (1967), 755–758; English transl. in Soviet Math. Dokl. **8** (1967).
3. _____, *The problem of Mal'tsev*, Sibirsk. Mat. Zh. **10** (1969), no. 5, 965–1005; English transl. in Siberian Math. J. **10** (1969).
4. _____, *On some questions of group and ring theory*, Dissertation, Sibirsk. Otdel. Akad. Nauk SSSR, Inst. Mat., Novosibirsk, 1969. (Russian)
5. _____, *Imbeddings into simple associative algebras*, Algebra i Logika **15** (1976), no. 2, 117–142; English transl. in Algebra and Logic **15** (1976).
6. N. Bourbaki, *Algèbre Commutative*, Hermann, Paris, 1961.
7. V. E. Govorov, *Graded algebras*, Mat. Zametki **12** (1972), no. 2, 197–204; English transl. in Math. Notes **12** (1972).
8. E. S. Golod, *On nil-algebras and residually finite p-groups*, Izv. Akad. Nauk SSSR Ser. Mat. **28** (1964), no. 2, 273–276; English transl. in Amer. Math. Soc. Transl. Ser. 2 **48** (1965).
9. P. M. Cohn, *Free Rings and Their Relations*, 2nd ed., Academic Press, London, 1985.
10. A. I. Kostrikin and I. R. Shafarevich, *Groups of homologies of nilpotent algebras*, Dokl. Akad. Nauk SSSR **115** (1957), 1066–1069. (Russian)
11. A. I. Mal'cev, *On the immersion of algebraic rings into a field*, Math. Ann. **113** (1937), 686–691.
12. _____, *On the immersion of associative systems into groups* I, Mat. Sb. **6** (1939), no. 2, 331–336. (Russian)
13. _____, *On the immersion of associative systems into groups* II, Mat. Sb. **8** (1940), no. 2, 251–264. (Russian)
14. V. A. Ufnarovskiĭ, *Poincaré series of graded algebras*, Mat. Zametki **27** (1980), no. 1, 21–32; English transl. in Math. Notes **27** (1980).
15. A. I. Shirshov, *Some algorithm problems for Lie algebras*, Sibirsk. Mat. Zh. **3** (1962), no. 2, 292–296. (Russian)
16. V. N. Gerasimov, *On the word problem for associative algebras with a single relation* III, All-Union Conference in Mathematical Logics, Novosibirsk, 1974. (Russian)
17. _____, *Distributive lattices of subspaces and the word problem for one-relator algebras*, Algebra i Logika **15** (1976), no. 4, 384–435; English transl. in Algebra and Logic **15** (1976).
18. _____, *Inverting ring homomorphisms*, XV All-Union Conference in Algebra, Krasnoyarsk, 1979. (Russian)
19. _____, *Inverting homomorphisms of rings*, Algebra i Logika **18** (1979), no. 6, 648–663; English transl. in Algeba and Logic **18** (1979).
20. J. Backelin, *La série de Poincaré-Betti d'une algèbre graduée de type fini à une relation est rationelle*, C. R. Acad. Sci. Paris Sér A **287** (1978), 843–846.
21. G. M. Bergman, *Commuting elements in free algebras and related topics in ring theory*, Thesis, Harvard Univ., Cambridge, MA, 1967.
22. _____, *Coproducts and some universal ring constructions*, Trans. Amer. Math. Soc. **200** (1974), 33–88.
23. A. J. Bowtell, *On a question of Mal'cev*, J. Algebra **7** (1967), 126–139.
24. P. M. Cohn, *Some remarks on the invariant basis property*, Topology **5** (1966), 215–228.
25. _____, *The embedding of firs in skew fields*, Proc. London Math. Soc. (3) **23** (1971), 193–213.
26. A. A. Klein, *Rings nonembeddable in fields with multiplicative semi-groups embeddable in groups*, J. Algebra **7** (1967), 100–125.

27. J. Lewin and T. Lewin, *On ideals of free associative algebras generated by a single element*, J. Algebra **8** (1968), 248–255.
28. S. Mac Lane, *Categories for the Working Mathematician*, Springer-Verlag, New York and Berlin, 1972.
29. O. Ore, *Linear equations in non-commutative fields*, Ann. of Math. (2) **32** (1931), 463–477.
30. _____, *Theory of non-commutative polynomials*, Ann. of Math. (2) **34** (1932), 480–508.
31. J. Shearer, *A graded algebra with a nonrational Hilbert series*, J. Algebra **62** (1980), 228–231.

Translated by MIRA BERNSTEIN

Part II

Representations of Algebras by Triangular Matrices

N. G. Nesterenko

Conventions and Notation

The word "algebra" means an associative algebra over a fixed commutative ring Φ with identity (the existence of an identity in an algebra is not assumed). All tensor products are considered over Φ. When we say that K is an algebra with identity, we assume that all K-modules are unital. A commutative regular ring means a commutative von Neumann regular ring with identity. We assume that only a finite number of summands differ from zero in all sums.

For a category $\mathscr{D}$ the class of its objects is denoted by $\text{ob}\,\mathscr{D}$, and the set of all morphisms from an object i into j is denoted by $\mathscr{D}(i, j)$. A small category $\mathscr{D}$ is called a category over Φ or a Φ-category if the set $\mathscr{D}(i, j)$ is a module over Φ for any two objects i and j, and multiplication by every fixed morphism is a linear mapping of the corresponding modules.

An algebra is called a T_n^M-algebra ($M \subseteq \{1, \ldots, n\}$) if it is embeddable in the algebra of upper triangular matrices of order n (over a commutative algebra), whose (i, i)th component equals zero for $i \notin M$. If $M = \{1, \ldots, n\}$, we write "T_n-algebra" instead of "T_n^M-algebra".

Also, we use the following notation:

$$[x, y] = xy - yx,$$

$$[x_1, \ldots, x_{n+1}] = [[x_1, \ldots, x_n], x_{n+1}];$$

$R^\# = R \oplus \Phi \cdot 1$ is the algebra R with an externally adjoined identity;

$R[X]$ is the algebra of polynomials (with a constant term) over an algebra R with the set X of commuting variables;

$R\langle X \rangle$ is the algebra of generalized polynomials (with a constant term) over an algebra R with the set X of noncommuting variables; in particular, $\Phi\langle X \rangle$ is the free algebra with identity generated by the set X;

$\Phi\langle X \rangle^{\cdot}$ is the algebra without identity freely generated by the set X;

$T_n(K)$ is the algebra of upper triangular matrices of order n over an algebra K;

(R, R) is the (commutator) ideal of an algebra R generated by all elements of the form $[a, b]$ $(a, b \in R)$.

Introduction

This dissertation concerns the theory of representations (embeddings) of associative algebras by matrices over a commutative algebra. Here the emphasis is on representations of algebras by triangular matrices of fixed order.

The fundamentals of this theory were laid out in Mal′tsev's work [9] of 1943. Later this theory attracted the attention of a good many mathematicians, and interest in it is not weakening even now. Research was done in various directions, one of which is closely connected with the theory of PI-algebras (algebras with a nontrivial identity). Once there even existed an opinion that all PI-algebras are representable, that is, they are exhausted by matrix subalgebras. Kaplansky's problem ([20] and [21], Problem 4) is connected with that. However, when V. N. Latyshev and P. M. Cohn answered this question in the negative, it became clear that by far not all PI-algebras are representable. At the present time, due to works of Anan′in in [1] and [2], representable varieties of algebras over a field of characteristic zero have been described in full. Locally representable varieties (that is, the varieties whose every finitely generated algebra is representable) have been described as well. In the case of algebras over an infinite field this result is due to Anan′in [3], while in the case of algebras over a finite field it is due to Kublanovskiĭ [7].

We mention also a direction connected with representability of relatively free (rf) algebras (that is, algebras free in a certain variety). The first result on the representability of rf algebras was obtained by Razmyslov [15] who proved that every rf algebra (with identity over a field of characteristic zero) that satisfies the identities of the algebra of matrices of the second order is embeddable in an algebra that is finitely generated as a module over its center. Beĭdar [4] proved that the algebra of endomorphisms of a finitely generated module over a commutative Noetherian algebra over a field is representable. It follows from these two results that every rf algebra (with identity over a field of characteristic zero) that satisfies the identities of the algebra of matrices of order two is representable.

Recent results of A. R. Kemer, who proved that every rf finitely generated algebra over an infinite field is representable, are of great importance.

We mention also a recent result of Procesi [23] that an algebra (with identity over a field of characteristic zero) with a trace is embeddable with the trace preserved into the algebra of matrices of order n over a commutative algebra if it satisfies the Hamilton-Cayley identity of degree n.

The problem of matrix representation of algebras is closely connected with the problem of representation of Φ-categories. A category $\mathscr{D}$ over Φ is called representable if there exists a Φ-linear univalent functor from this category into a commutative algebra with identity considered as a category with a single object. In the case when $\mathrm{ob}\mathscr{D} = \{1, \ldots, n\}$, ordinary "matrix" operations on the module $\bigoplus \mathscr{D}(i, j)$ $(i, j \in \mathrm{ob}\mathscr{D})$ determine a structure of a matrix algebra $M(\mathscr{D})$ over $\mathscr{D}$. It is clear that in this case the representability of the category $\mathscr{D}$ means the existence of an embedding $\varphi: M(\mathscr{D}) \to T_n(K)$ of algebras with identity (K is a commutative algebra) that preserves the components (that is, transforms a matrix with the zero (i, j)th component into a matrix with the zero (i, j)th component). Thus, the problem of matrix representability of algebras is reduced to the problem of embedding algebras into the matrix algebra over a representable category.

A category $\mathscr{D}$ over Φ with a partial order on the set $\mathrm{ob}\mathscr{D}$ such that $\mathscr{D}(i, j) \neq 0 \Rightarrow i \leq j$, is called triangular (respectively, unitriangular) if for every object i from the category $\mathscr{D}$ the algebra $\mathscr{D}(i, i)$ is commutative (is isomorphic to Φ).

In the first chapter of this dissertation representable triangular categories are studied. One of the first in this direction is Bergman's result [17] that each unitriangular category is representable.

Proposition 1 contains the principal technical criterion for the representability of a triangular category. By means of this criterion Theorem 1 gives a description of representable triangular categories in the language of quasi-identities. The following criterion (see Corollary 1) that gives sufficient conditions for the representability of triangular categories turns out to be useful for applications: a triangular category $\mathscr{D}$ over a field is representable if there exist homomorphisms $\chi_{ij}: \mathscr{D}(i, i) \to \mathscr{D}(j, j)$ $(j < i)$ of algebras with identity such that $ab = \chi_{ij}(b)a$ for any morphisms $a \in \mathscr{D}(j, i)$, $b \in \mathscr{D}(i, i)$.

Bergman [17] formulated (in the language of quasi-identities) certain conditions valid for all representable Φ-categories. In the same paper Bergman raised the problem of sufficiency of these conditions for the representability of the category. Section 2 of Chapter 1 of this dissertation contains an example that answers this question in the negative even for triangular categories that contain more than three objects.

In the final section of the first chapter we study embeddings (that preserve grading) of graded algebras of the form $R = R_0 \oplus \cdots \oplus R_{n-1}$ into algebras of upper triangular $n \times n$-matrices over a commutative algebra (with the natural diagonal grading). Theorem 1 makes it possible to give a criterion (in the language of quasi-identities) for algebras over a field that is necessary and sufficient for the existence of such an embedding (see Corollary 3).

Many of the works devoted to representation of algebras by matrix algebras over a commutative algebra really consider representations by unitriangular matrices. Also, as observed by Anan'in [2], for characteristic zero the class of representable varieties does not change if representability by square matrices is replaced by representability by upper triangular matrices. In this connection there arose the problem of the description of T_n-algebras. Lewin [22] proved that a T_2-algebra (over a field) is an algebra that satisfies the identity $[x, y][z, t] = 0$. In the case of algebras over an arbitrary commutative ring T_2-algebras were described in [11]. Note that a base for identities of the algebra of upper triangular matrices was found by Yu. N. Mal'tsev [10].

In the middle of the seventies Anan'in put forth the conjecture that an algebra R (over a field) is embeddable in the algebra of upper triangular matrices of order n over a commutative algebra if and only if it satisfies the following condition:

If A is a finite set of strings of the form $\alpha = (a_1, x_1, \ldots, a_n, x_n, a_{n+1})$ ($a_i \in R^\#$, $x_j \in R$) and, for every nonempty subset $\{i_1, \ldots, i_k\} \subseteq \{1, \ldots, n\}$ and arbitrary elements $y_{i_1}, \ldots, y_{i_k} \in R$ the equality

$$\sum_{\alpha \in A} a_1 x_1 a_2 x_2 \cdots a_{i_1} \delta_{x_{i_1} y_{i_1}} a_{i_1+1} \cdots a_{i_k} \delta_{x_{i_k} y_{i_k}} a_{i_k+1} \cdots a_n x_n a_{n+1} = 0,$$

holds, where δ_{xy} is the Kronecker symbol, then the equality

$$\sum_{\alpha \in A} a_1 x_1 a_2 x_2 \cdots a_n x_n a_{n+1} = 0$$

holds in the algebra R.

Anan'in was the first to observe that this condition holds in every T_n-algebra.

For $n = 3$ this hypothesis takes a simpler form: an algebra R is a T_3-algebra if and only if

$$\sum_i a_i b_i = 0 \Rightarrow \sum_i a_i x b_i = 0$$

for all $a_i, b_i \in (R, R)$, $x \in R$.

Anan'in suggested this conjecture wishing above all to prove the following theorem (which he proved later by other means):

THEOREM (Anan'in [2]). *An algebra R over a field that satisfies the identities*

$$[x_1, \ldots, x_t]y_1 \cdots y_t[z_1, \ldots, z_t] = 0, \qquad [x_1, y_1] \cdots [x_t, y_t] = 0,$$

is a T_n-algebra, where n depends on t only.

The main result of the second chapter is Theorem 2 that gives necessary and sufficient conditions (analogous to Anan'in's quasi-identities) for the existence of an embedding $\varphi: R \to T_n(K)$, where K is a commutative algebra and $\varphi(r) = (\varphi_{ij}(r))_{i \leq j}$, such that $\ker \varphi_{ii} = I_i$ ($i = 1, \ldots, n$) are fixed ideals of the algebra R that contain its commutative ideal. Here the proof is given in the language of generalized polynomials, in which Anan'in's conjecture takes the following form:

An algebra R is a T_n-algebra if and only if, for every n-polynomial f over R, the quasi-identity

$$\bigwedge_{\substack{x_1, \ldots, x_k \in X \\ k > 0}} \left(\frac{\partial^k f}{\partial x_1 \cdots \partial x_k} = 0 \right) \Rightarrow f = 0 \qquad (*)$$

holds.

An n-polynomial is defined as a generalized polynomial over an algebra R with variables from the set $X = X_1 \dot\cup \cdots \dot\cup X_n$ (X_i is a countable set) that is decomposable into a sum of monomials of the form $a_1 x_1 a_2 \cdots x_n a_{n+1}$, where $a_i \in R^\#$, $x_j \in X_j$.

Particular cases of Theorem 2, when each of the ideals $I_1, \ldots, I_n$ coincides either with the entire algebra or with its commutator ideal, are of major interest.

THEOREM 3. *An algebra R over a commutative regular ring is a T_n^M-algebra ($M \subseteq \{1, \ldots, n\}$) if and only if the quasi-identity $(*)$ holds for every (n, M)-polynomial f over R.*

Here an (n, M)-polynomial is defined as a generalized polynomial over R that is obtained from an n-polynomial by means of replacing all variables from the sets X_i ($i \notin M$) by suitable elements from R.

Setting $M = \{1, \ldots, n\}$ in this theorem, we obtain a positive answer to Anan'in's conjecture. On the other hand, assuming that $M = \varnothing$, we obtain a known (in the case of algebras over a field) result, proved independently by L'vov [8] and Bergman and Vovsi [18], that a nilpotent of index n algebra can be embedded in an algebra of strictly upper triangular matrices of order n over a commutative algebra.

The criterion of representability of a triangular category given above plays an important role in the proof of Theorem 2. Also, algebras with additional structure appear—algebras with diagonal. We consider the definition of algebras with diagonal in more detail. It is easy to show that the identity

$$(x_1 - d_1(x_1))(x_2 - d_2(x_2)) \cdots (x_n - d_n(x_n)) = 0$$

(with a diagonal of order n) holds in the algebra $T_n(K)$ (K is a commutative algebra), where d_k ($1 \leq k \leq n$) is a diagonal mapping, that is, the homomorphism of the algebra $T_n(K)$ into its center that maps a matrix (x_{ij}) into the scalar matrix $x_{kk} E_n$. Clearly, $d_i d_j = d_j$. It turns out that the identity with diagonal plays the same role for the algebra of upper triangular matrices as the Hamilton-Cayley identity for the full matrix algebra. In particular, for every algebra over a commutative regular ring with diagonal D of order n (that is, with the string $(D_1, \ldots, D_n)$ of homomorphisms of the algebra R into its center such that $D_i D_j = D_j$), that satisfies the identity with diagonal, there exists an embedding $\varphi \colon R \to T_n(K)$ (K is a commutative algebra). Although in the general case this embedding does not preserve diagonal mappings, that is, $\varphi D_i \neq d_i \varphi$, we have, however, the following statements:

THEOREM 4. *Every algebra R with diagonal of order $n < 4$ that satisfies the identity with diagonal of order n is embeddable with preservation of the diagonal into the algebra $T_n(K)$ for a suitable commutative algebra K.*

THEOREM 5. *Let R be an algebra with identity over a commutative regular ring with diagonal $(d_1, \ldots, d_n)$ that satisfies the identity with diagonal, and the diagonal mappings $d_1, \ldots, d_n$ preserve the identity and act into the subalgebra generated by the identity. Then R can be embedded with preservation of the diagonal into the algebra $T_n(K)$ for a suitable commutative algebra K.*

Observe that Theorem 4 is no longer true already for $n = 4$, and that the regularity of the ground ring cannot be dropped from Theorem 5 (corresponding examples are given in the final section of the second chapter). Also, in the final section of the second chapter we produce examples that answer in the negative certain questions raised by L'vov and Anan'in and connected with embedding algebras in algebras of matrices of a fixed order and form.

The author is indebted to I. V. L'vov for a short method of reducing proofs of statements of the type of Anan'in's conjecture from algebras over a field to algebras over a regular commutative ring with identity.

In the third chapter we consider special representations of graded algebras (over an arbitrary commutative ring) all of whose homogeneous components except maybe from the first to the nth one equal zero. In Proposition 5 we give a necessary condition for embeddability of such an algebra in a subalgebra of the algebra $T_n(K)$ (K is a commutative algebra) of a special form. As a corollary we obtain a known result of Bergman [17] that every graded algebra, all of whose homogeneous components except maybe from the first to the $(n-1)$st one equal zero, is embeddable in the algebra of strictly upper triangular matrices of order n over a commutative algebra. In addition, we prove that every $T_{n+2}^{\varnothing}$-algebra is a $T_n^{\{1,n\}}$-algebra for $n > 1$ and it is a $T_{n+1}^{\{1\}}$-algebra for $n > 0$. In particular, it follows that a nilpotent of index $n > 3$ algebra over a field is representable as an algebra of triangular matrices of order $n - 2$ over a commutative algebra. On the other hand, in §2 of the third chapter we produce an example of a nilpotent of index $n > 3$ algebra that is not embeddable in an algebra of full matrices of order $< n - 2$ over a commutative ring. In the case of algebras over an arbitrary commutative ring with identity we obtain (using Bergman's result [19] that a nilpotent of index $n + 1$ algebra is embeddable in a graded algebra all of whose homogeneous components except maybe from the first to the 2^{n-1}th one equal zero) that a nilpotent of index $n + 1$ algebra is always embeddable in the algebra of upper triangular matrices of order $2^{n-1} - 1$ over a commutative algebra.

All the results were presented at different times at the seminar in the theory of rings of the Institute of Mathematics of the Siberian Branch of the Academy of Sciences of the USSR. Some of them were presented at the seminar "Algebra and Logic" in the Novosibirsk State University, at the seminar in general algebra at the Moscow State University and at the XIXth All-Union algebraic conference in L'vov in 1987.

The author expresses his sincere gratitude to his scientific advisor professor L. A. Bokut' and also to I. V. L'vov and A. Z. Anan'in, scientific contacts with whom facilitated in many respects the appearance of these results.

Chapter 1
Representability of Triangular Categories and Graded Algebras

In this chapter we obtain, in the language of categories, some statements on the representability of graded algebras, which permit us to describe graded T_n-algebras and turn out to be useful for study of T_n-algebras. Here ideas of Bergman's paper [17] are used and developed, and an example is given that answers in the negative one of his questions on representable categories.

§1. Preliminary remarks on representable triangular categories

If $\mathscr{D}$ and $\mathscr{C}$ are two Φ-categories, then, when speaking of functors from $\mathscr{D}$ into $\mathscr{C}$, we always assume them to be Φ-functors, that is, additive (covariant) functors permutable with the elements of the ring Φ.

Following [16] a functor $F: \mathscr{D} \rightsquigarrow \mathscr{C}$ is called univalent if the restriction of F to the set $\mathscr{D}(i, j)$ is injective for any two objects i and j.

Omitting all axioms related to identity morphisms in the definition of a category and functor we obtain a concept of a precategory and a functor of precategories, respectively. A Φ-precategory, a univalent functor of precategories, etc. are defined in a natural way.

If $\mathscr{D}$ is a precategory over Φ, then the usual "matrix" operations on $\bigoplus \mathscr{D}(i, j)$ $(i, j \in \mathrm{ob}\mathscr{D})$ determine a matrix algebra $M(\mathscr{D})$ over $\mathscr{D}$. The correspondence $\mathscr{D} \mapsto M(\mathscr{D})$ makes it possible to transfer ring concepts to the case of categories. For example, a precategory $\mathscr{S}$ over Φ is called an ideal of a Φ-category $\mathscr{D}$ if $\mathrm{ob}\mathscr{S} = \mathrm{ob}\mathscr{D}$, $\mathscr{S}(i, j) \subseteq \mathscr{D}(i, j)$ $(i, j \in \mathrm{ob}\mathscr{S})$ and $M(\mathscr{S})$ is an ideal of the algebra $M(\mathscr{D})$. Analogously one defines a factor category $\mathscr{D}/\mathscr{S}$; here the algebra $M(\mathscr{D}/\mathscr{S})$ is naturally isomorphic to the factor algebra $M(\mathscr{D})/M(\mathscr{S})$.

If $\mathscr{D}$ is a Φ-category and K a commutative algebra with identity, then $K \otimes \mathscr{D}$ denotes the K-hull of the category $\mathscr{D}$, that is, a K-category whose objects are objects of the category $\mathscr{D}$ and the set of morphisms from an object i into j coincides with the module $K \otimes \mathscr{D}(i, j)$. The functor from $\mathscr{D}$ into $K \otimes \mathscr{D}$ that preserves objects and transfers a morphism a into a morphism $1 \otimes a$ is called canonical. Since tensor product commutes with direct sums, we obtain $M(K \otimes \mathscr{D}) \approx K \otimes M(\mathscr{D})$.

For a Φ-category $\mathscr{D}$, let $K_{\mathscr{D}}$ denote the tensor product of the algebras $\mathscr{D}(k, k)$ ($k \in \mathrm{ob}\,\mathscr{D}$) (see [6, p. 456] for a definition of tensor product of an infinite number of algebras with identity). Here we agree to identify the algebras $\mathscr{D}(k, k)$ ($k \in \mathrm{ob}\,\mathscr{D}$) with their natural images in $K_{\mathscr{D}}$.

A (pre)category over Φ is called representable if there exists a univalent functor from this (pre)category into a commutative algebra with identity considered as a Φ-category with a single object.

A (pre)category $\mathscr{D}$ over Φ with a partial order on the set $\mathrm{ob}\,\mathscr{D}$ such that $\mathscr{D}(i, j) \neq 0 \Rightarrow i \neq j$ is called a triangular (pre)category over Φ if $\mathscr{D}(k, k)$ is a commutative algebra for any object k from $\mathscr{D}$. A triangular category $\mathscr{D}$ over Φ is called unitriangular if the algebra $\mathscr{D}(k, k)$ is isomorphic to Φ for every object k from $\mathscr{D}$.

In the sequel, unless stated otherwise, a triangular (pre)category means a triangular (pre)category over Φ.

For a triangular precategory $\mathscr{D}$, let $\mathscr{D}^{\#}$ denote the triangular category with the same objects as $\mathscr{D}$, where $\mathscr{D}^{\#}(i, j) = \mathscr{D}(i, j)$ for $i = j$ and $\mathscr{D}^{\#}(i, i)$ is the algebra $\mathscr{D}(i, i)$ with externally adjoined identity, that is, the identity morphism of the object i in the category $\mathscr{D}^{\#}$. Clearly, representability of the precategory $\mathscr{D}$ is equivalent to representability of the category $\mathscr{D}^{\#}$. Thus, the problem of representability of triangular precategories is reduced to the problem of representability of triangular categories.

In this direction Bergman's result [17] that every unitriangular category is representable was one of the first. In the same paper Bergman observed that it is necessary for the representability of a Φ-category $\mathscr{D}$ that, for every object k, the algebra $\mathscr{D}(k, k)$ be commutative and the canonical functor from the category $\mathscr{D}$ into its $K_{\mathscr{D}}$-hull be an embedding.

For a triangular category $\mathscr{D}$, let $\mathscr{D}_{*}$ denote its $K_{\mathscr{D}}$-hull and $P_{\mathscr{D}}$ the ideal of the $K_{\mathscr{D}}$-category $\mathscr{D}_{*}$ generated by all morphisms of the form

$$a \otimes uv - 1 \otimes uav, \qquad (1)$$

where $u \in \mathscr{D}(i, k)$, $a \in \mathscr{D}(k, k)$, $v \in \mathscr{D}(k, j)$, $i < k < j$. Observe that the module $P_{\mathscr{D}}(i, j)$ equals zero if the interval $\langle i, j \rangle := \{k \in \mathrm{ob}\,\mathscr{D} | i < k < j\}$ is empty.

In the sequel $[A \otimes B]$ denotes the natural image of the module $A \otimes B$ in $\mathscr{D}_{*}(i, j)$ for submodules $A \subseteq K_{\mathscr{D}}$, $B \subseteq \mathscr{D}(i, j)$.

PROPOSITION 1. *It is necessary and sufficient for the representability of a triangular category $\mathscr{D}$ that*
  1) *the canonical functor from $\mathscr{D}$ into its $K_{\mathscr{D}}$-hull be univalent and*
  2) *the identity*

$$[1 \otimes \mathscr{D}(i, j)] \cap P_{\mathscr{D}}(i, j) = 0$$

hold for any two objects i and j from $\mathscr{D}$ such that $\langle i, j \rangle \neq \varnothing$.

PROOF. Let φ denote the canonical functor from $\mathscr{D}$ into $\mathscr{D}_*$.

Necessity. Let F be a univalent Φ-functor from $\mathscr{D}$ into a commutative algebra B with identity considered as a category with a single object. Setting $a \cdot b = F(a)b$ for $a \in \mathscr{D}(k,k)$, $b \in B$, we endow B with a structure of a $K_{\mathscr{D}}$-category with a single object. Here the functor F induces a $K_{\mathscr{D}}$-functor $\tilde{F}: \mathscr{D}_* \rightsquigarrow B$ ($F = \tilde{F}\varphi$), that obviously annihilates all morphisms of the form (1). Both conditions 1) and 2) follow immediately.

Sufficiency. Let Ω be the ideal of the $K_{\mathscr{D}}$-category $\mathscr{D}_*$ that is generated by all morphisms of the form (1), where $u \in \mathscr{D}(i,k)$, $a \in \mathscr{D}(k,k)$, $v \in \mathscr{D}(k,j)$, $i \leq k \leq j$. Clearly, the correspondence $ab \otimes d \mapsto 1 \otimes adb$ ($d \in \mathscr{D}(i,j)$, $a \in \mathscr{D}(i,i)$, $b \in \mathscr{D}(j,j)$) induces an endomorphism of the A-module $\mathscr{D}_*(i,j)$, where A is the subalgebra of $K_{\mathscr{D}}$ generated by the natural images of the algebras $\mathscr{D}(k,k)$, where the object k differs from i and j. Clearly, this endomorphism maps $\Omega(i,j)$ onto $P_{\mathscr{D}}(i,j)$, and its restriction to the submodule $[1 \otimes \mathscr{D}(i,j)]$ is injective. Therefore, condition 2) implies $\Omega(i,j) \cap [1 \otimes \mathscr{D}(i,j)] = 0$. Thus, by condition 1), the category $\mathscr{D}$ is embeddable in the factor category $H = \mathscr{D}_*/\Omega$.

Observing that the epimorphism $\psi_k: K_{\mathscr{D}} \otimes \mathscr{D}(k,k) \to K_{\mathscr{D}}$ ($a \otimes c \mapsto ac$) of $K_{\mathscr{D}}$-algebras annihilates the ideal $\Omega(k,k)$ and taking into consideration the fact that $b - \psi_k(b) \otimes 1 \in \Omega(k,k)$ for arbitrary $b \in \mathscr{D}_*(k,k)$, we obtain an isomorphism $H(k,k) \to K_{\mathscr{D}}$ ($b + \Omega(k,k) \mapsto \psi_k(b)$, $b \in \mathscr{D}_*(k,k)$) of $K_{\mathscr{D}}$-algebras. It follows that H is a unitriangular $K_{\mathscr{D}}$-category. It remains to use Bergman's result [17] on the representability of unitriangular categories. Proposition 1 is proved.

Let $\mathscr{D}$ be a triangular (pre)category over Φ. A homomorphism $\chi: \mathscr{D}(i,i) \to \mathscr{D}(j,j)$ of algebras that preserves identity in the case of categories is called an (i,j)-*translation* if the equalities $ab = \chi(b)a$ and $bc = c\chi(b)$ hold for all morphisms $a \in \mathscr{D}(j,i)$, $c \in \mathscr{D}(i,j)$, $b \in \mathscr{D}(i,i)$.

We say that a triangular (pre)category possesses a *left (right) translation* if an (i,j)-translation exists for any two objects i and j such that $i > j$ (respectively, $i < j$).

LEMMA 1. *A triangular category $\mathscr{D}$ with a left (right) translation satisfies condition* 2) *of Proposition* 1.

PROOF. Without loss of generality, let a triangular category $\mathscr{D}$ have a left translation, that is, there exist (i,j)-translations $\chi_{i,j}$ for all $i > j$. Fix two objects k and l of the category $\mathscr{D}$ such that the interval $\langle k, l \rangle$ is not empty. Let A (respectively, B) denote the subalgebra of $K_{\mathscr{D}}$ generated by natural images of the algebras $\mathscr{D}(j,j)$ for which $j > k$ (respectively, $j \not> k$). Define a structure of an A-module on the module $\mathscr{D}(k,l)$ by setting $a \cdot d = \chi_{jk}(a)d$ for $a \in \mathscr{D}(j,j)$, $d \in \mathscr{D}(k,l)$, $j > k$. It is easy to see that the endomorphism $ba \otimes d \mapsto b \otimes a \cdot d$ ($a \in A$, $b \in B$, $d \in \mathscr{D}(k,l)$) of the module $\mathscr{D}_*(k,l)$ annihilates the submodule $P_{\mathscr{D}}(k,l)$ and acts identically on the submodule $[1 \otimes \mathscr{D}(k,l)]$. Lemma 1 is proved.

Clearly, the canonical functor from the triangular category $\mathscr{D} = \mathscr{S}^{\#}$ ($\mathscr{S}$ is a triangular precategory) into its $K_{\mathscr{D}}$-hull is univalent. Also, every (i, j)-translation of the precategory $\mathscr{S}$ can be naturally extended to an (i, j)-translation of the category $\mathscr{D}$. Thus, Lemma 1 and Proposition 1 imply

COROLLARY 1. 1) *Every triangular precategory with a left (right) translation is representable.*

2) *Every triangular category over a field that possesses a left (right) translation is representable.*

Observe that, for a triangular (pre)category with a finite set of objects, the existence of a left translation follows from the existence of (i, j)-translations for all objects i and j such that $i = j + 1$ is the successor of the element j.

§2. Quasi-identities of representable triangular categories

In his article [17] Bergman formulated, in the language of quasi-identities, certain conditions satisfied in all representable Φ-categories. There Bergman raised a problem of the sufficiency of these conditions for representability of the category. In the case of a unitriangular category $\mathscr{D}$ over a field Bergman's conditions take the form:

for every $a_p \in \mathscr{D}(i, j)$, $b \in \mathscr{D}(j, j)$, $c_p \in \mathscr{D}(j, k)$ $(i < j < k)$

$$\sum_p a_p c_p = 0 \Rightarrow \sum_p a_p b c_p = 0. \qquad (2)$$

The following proposition shows that this condition is sufficient for representability of a triangular category over a field in which, for any two objects i and j, there exists at most one object k lying between them. Further, we produce an example showing that condition (2) is not sufficient for representability of a triangular category over a field if its set of objects contains an interval $\langle i, j \rangle$ that consists of more than a single element.

PROPOSITION 2. *Let a triangular category $\mathscr{D}$ satisfy the following conditions:*

1) *the canonical functor from $\mathscr{D}$ into $\mathscr{D}_*$ is univalent;*
2) *condition (2) holds;*
3) *the partial order on the set $\mathrm{ob}\,\mathscr{D}$ induces a linear order on the interval $\langle i, j \rangle$ for any objects i and j of $\mathscr{D}$;*
4) *if for objects i and j of $\mathscr{D}$ the interval $\langle i, j \rangle$ contains more than a single object, then*

$$\mathscr{D}(i, j) = \mathscr{D}(i, i_1)\mathscr{D}(i_1, i_2) \cdots \mathscr{D}(i_t, j)$$

for every finite chain $i < i_1 < \cdots < i_t < j$ of objects of the category $\mathscr{D}$;

5) *if for objects i and j of $\mathscr{D}$ the interval $\langle i, j \rangle$ contains exactly one object k, then*

$$[K_{\mathscr{D}} \otimes \mathscr{D}(i, k)\mathscr{D}(k, j)] \cap [1 \otimes \mathscr{D}(i, j)] = [1 \otimes \mathscr{D}(i, k)\mathscr{D}(k, j)].$$

Then the category $\mathscr{D}$ is representable.

REMARK 1. As we have already observed, conditions 1) and 2) are necessary for the representability of a category. Besides, condition 5) certainly holds in the case when the ground ring is regular. Conditions 3) and 4) hold trivially for categories that contain at most three objects.

PROOF OF PROPOSITION 2. It suffices to verify that the category $\mathscr{D}$ satisfies condition 2) of Proposition 1. Let i and j be objects of $\mathscr{D}$ such that $i < j$. Using the implication (2) define an endomorphism $\overline{b}$ for an arbitrary object $k \in \langle i, j \rangle$ and a morphism $b \in \mathscr{D}(k, k)$ by setting

$$\overline{b}\left(\sum_p a_p c_p\right) = \sum_p a_p b c_p \quad (a_p \in \mathscr{D}(i, k),\ c_p \in \mathscr{D}(k, j)).$$

Take an arbitrary element $x \in P_{\mathscr{D}}(i, j) \cap [1 \otimes \mathscr{D}(i, j)]$. Clearly, it is representable in the form of a $K_{\mathscr{D}}$-linear combination of morphisms of the form (1), where $k \in \{i_1, \ldots, i_t\} \subseteq \langle i, j \rangle$. By (3), we may assume that $i_1 < \cdots < i_t$. Observe that

$$P_{\mathscr{D}}(i, j) \subseteq [K_{\mathscr{D}} \otimes \mathscr{D}(i, i_1)\mathscr{D}(i_1, i_2) \cdots \mathscr{D}(i_t, j)].$$

Indeed, if the interval $\langle i, j \rangle$ contains more than a single element, this inclusion follows from condition 4). If this interval contains at most one element, then the desired inclusion follows directly from the definition of the ideal $P_{\mathscr{D}}$.

Define an endomorphism φ of the submodule

$$[K_{\mathscr{D}} \otimes (\mathscr{D}(i, i_1) \cdot \mathscr{D}(i_1, i_2) \cdots \mathscr{D}(i_t, j))],$$

by $\varphi(aa_1 \cdots a_t \otimes d) = a \otimes \overline{a}_1 \cdots \overline{a}_t(d)$, where $d \in \mathscr{D}(i, i_1) \cdot \mathscr{D}(i_1, i_2) \cdots \mathscr{D}(i_t, j)$, $a_p \in \mathscr{D}(i_p, i_p)$, and a is an element of the subalgebra of $K_{\mathscr{D}}$ generated by the natural images of algebras $\mathscr{D}(k, k)$, where $k \notin \{i_1, \ldots, i_t\}$. Clearly, φ annihilates x and acts identically on the submodule

$$[1 \otimes \mathscr{D}(i, i_1)\mathscr{D}(i_1, i_2) \cdots \mathscr{D}(i_t, j)].$$

Therefore, $x = 0$. Proposition 2 is proved.

In particular, it follows from Proposition 2 that an arbitrary triangular precategory whose set of objects does not contain chains of length three is representable.

It is easy to verify that for every representable triangular category $\mathscr{D}$ that contains objects i, k, l, and j such that $i < k, l < j$ the following condition holds:

Let $\mathfrak{X} \subseteq \mathscr{D}(k, k)$, $\mathfrak{Y} \subseteq \mathscr{D}(l, l)$ be finite subsets and, for all $x \in \mathfrak{X}$, $y \in \mathfrak{Y}$, there are given finite subsets $A_{xy}, A_x \subseteq \mathscr{D}(i, k) \times \mathscr{D}(k, j)$, B_{xy},

$B_y \subseteq \mathscr{D}(i, l) \times \mathscr{D}(l, j)$. Assume that all equalities of the form

$$\sum_{(u,v) \in A_{xy}} uv = \sum_{(u,v) \in B_{xy}} uv, \quad \sum_{x \in \mathfrak{X}} \sum_{(u,v) \in A_{xy}} uxv = \sum_{(u,v) \in B_y} uv \qquad (3)$$

$$\sum_{y \in \mathfrak{Y}} \sum_{(u,v) \in B_{xy}} uyv = \sum_{(u,v) \in A_x} uv$$

hold. Then the equality

$$\sum_{x \in \mathfrak{X}} \sum_{(u,v) \in A_x} uxv = \sum_{y \in \mathfrak{Y}} \sum_{(u,v) \in B_y} uyv$$

holds.

It is not difficult to produce now an example of a nonrepresentable triangular category $\mathscr{D}$ over a field Φ that satisfies condition (2). Set

$\mathrm{ob}\mathscr{D} = \{1, 2, 3, 4\}$, $\mathscr{D}(i, j) = 0 \ (i > j)$, $\mathscr{D}(1, 1) \approx \Phi$,
$\mathscr{D}(4, 4) \approx \Phi$, $\mathscr{D}(2, 2) = \Phi \oplus \Phi x_1$, $\mathscr{D}(3, 3) = \Phi \oplus \Phi x_2$,
$\mathscr{D}(2, 3) = 0$, $\mathscr{D}(1, 2) = \Phi x_3 \otimes \mathscr{D}(2, 2)$, $\mathscr{D}(1, 3) = \Phi x_6 \otimes \mathscr{D}(3, 3)$,
$\mathscr{D}(2, 4) = \mathscr{D}(2, 2) \otimes (\Phi x_4 + \Phi x_5)$, $\mathscr{D}(3, 4) = \mathscr{D}(3, 3) \otimes \Phi x_7$,
$\mathscr{D}(1, 4) = \Phi x_8 + \Phi x_9 + \Phi x_{10}$.

Here $x_1, x_2, \ldots, x_{10}$ are linearly independent elements of a certain space. The composition of morphisms in the category $\mathscr{D}$ is defined by the equalities

$x_1^2 = 0$, $x_2^2 = 0$, $x_3 x_1 = x_3 \otimes x_1$, $x_1 x_4 = x_1 \otimes x_4$,
$x_1 x_5 = x_1 \otimes x_5$, $x_2 x_7 = x_2 \otimes x_7$, $x_6 x_2 = x_6 \otimes x_2$, $x_3 x_1 x_4 = 0$,
$x_3 x_1 x_5 = x_{10}$, $x_3 x_4 = x_6 x_7 = x_8$, $x_3 x_5 = x_6 x_2 x_7 = x_9$.

It is not difficult to verify that the category $\mathscr{D}$ so obtained satisfies condition (2). On the other hand, if it were representable, then by (3), it would have followed from the equalities $x_3 x_4 = x_6 x_7$, $x_3 x_1 x_4 = 0$, $x_6 x_2 x_7 = x_3 x_5$ that $x_3 x_1 x_5 = 0$. However, $x_3 x_1 x_5 = x_{10} \neq 0$.

Before stating the theorem that gives a representability criterion for a triangular category $\mathscr{D}$ in the language of quasi-identities, we introduce certain definitions and notations. Every finite subset f of the Cartesian product $\mathscr{D}(i, k) \times \mathscr{D}(k, j)$ $(i < k < j)$ is called an (i, k, j)-set (over $\mathscr{D}$). Here we set

$$D_a f = \sum_{(u,v) \in f} uav,$$

for the morphism $a \in D(k, k)$.

An element $(x_i)_{i \in I}$ of the Cartesian product $\prod_{i \in I} \mathscr{D}(i, i)$ $(I \subseteq \mathrm{ob}\mathscr{D})$ is denoted by x_I. The subset of all such elements for which $x_i = 1$ for all objects $i \in I$ except, possibly, for a finite number of them is denoted by D_I; here we set $1_I = (1_i)_{i \in I}$.

THEOREM 1. *For the representability of a triangular category $\mathscr{D}$ it is necessary and, if the module $\mathscr{D}(i, i)$ is free and has a basis that contains the identity morphism 1_i for every object i of $\mathscr{D}$, then it is also sufficient, that for any two objects i and j of $\mathscr{D}$ the following condition hold:*
(**) *Let all equalities of the form*

$$\sum_{k \in I} D_{1_k} f_{k, x_I} = 0, \qquad \sum_{y_r \in \mathscr{D}(r, r)} D_{y_r} f_{r, y_I} = 0,$$

where $x_I, y_I \in \mathscr{D}_I$, $x_I \neq 1_I$, $r \in I$, hold for all (i, p, j)-sets f_{p, x_I} ($I = \langle i, j \rangle$, $p \in I$, $x_I \in \mathscr{D}_I$), among which only finitely many are nonempty. Then the equality

$$\sum_{k \in I} D_{1_k} f_{k, 1_I} = 0$$

holds.

PROOF. *Necessity.* Let F be a univalent Φ-functor from $\mathscr{D}$ into a commutative algebra with identity considered as a category with a single object and let the antecedents of condition (**) hold for the objects i and j of the category $\mathscr{D}$. Set

$$g_{p, x_I} = F(D_{1_p}, f_{p, x_I}) \qquad (p \in I, \ x_I \in \mathscr{D}_I).$$

Taking into consideration that

$$F(D_{y_r} f_{r, y_I}) = F(y_r) g_{r, y_I},$$

from the antecedents of condition (**) we obtain

$$\sum_{k \in I} g_{k, x_I} = 0, \qquad \sum_{y_r \in \mathscr{D}(r, r)} F(y_r) g_{r, y_I} = 0,$$

where $x_I, y_I \in \mathscr{D}_I$, $x_I \neq 1_I$, $r \in I$. It follows that

$$\sum_{k \in I} g_{k, 1_I} = \sum_{x_I \in \mathscr{D}_I} \left\{ \prod_{i \in I} F(x_i) \right\} \sum_{k \in I} g_{k, x_I}$$

$$= \sum_{k \in I} \sum_{x_{I_k} \in \mathscr{D}_{I_k}} \left\{ \prod_{i \in I_k} F(x_i) \right\} \sum_{x_k \in \mathscr{D}(k, k)} F(x_k) g_{k, x_I} = 0,$$

where $I_k = I \setminus \{k\}$. Since the functor F is univalent, we obtain the desired equality.

Sufficiency. Since each module $\mathscr{D}(k, k)$ possesses a basis $\mathfrak{M}_k$ containing the morphism 1_k, the canonical functor from $\mathscr{D}$ into $\mathscr{D}_*$ is univalent. Therefore, it suffices to verify only condition 2) of Proposition 1.

Let i and j be objects from $\mathscr{D}$, where the interval $I = \langle i, j \rangle$ is not empty. Let V denote the set of all elements of $\mathscr{D}_*(i, j)$ of the form

$$\sum_{x_I \in \mathscr{D}_I} \left\{ \prod_{i \in I} x_i \right\} \otimes \sum_{k \in I} D_{1_k} f(k, x_I), \qquad (4)$$

where $f(k, x_I)$ is an (i, k, j)-set (nonempty only for a finite number of strings of the form (k, x_I) for which $k \in I$, $x_k \in \mathfrak{M}_k$) and let all equalities of the form

$$\sum_{y_r \in \mathscr{D}(r,r)} D_{y_r} f_{(r, y_I)} = 0 \qquad (r \in I, y_I \in \mathscr{D}_I) \tag{5}$$

hold.

Obviously, the set V is a Φ-submodule in $\mathscr{D}_*(i, j)$. Since $ba \otimes uv - b \otimes uav = (ba \otimes uv - 1 \otimes ubav) - (b \otimes uav - 1 \otimes ubav)$, where $u \in \mathscr{D}(i, k)$, $v \in \mathscr{D}(k, j)$, $a, b \in \mathscr{D}(k, k)$, the submodule $P_\mathscr{D}(i, j)$ is generated as an abelian group by the elements of the form

$$\left(\prod_{i \in J} a_i\right) \otimes uv - \left(\prod_{i \in J_k} a_i\right) \otimes u a_k v,$$

where $u \in \mathscr{D}(i, k)$, $v \in \mathscr{D}(k, j)$, $a_J = (a_i)_{i \in J} \in \mathscr{D}_J$, $J = \mathrm{ob}\mathscr{D}$, $k \in I$, $J_k = J\setminus\{k\}$, and $a_i \in \mathfrak{M}_i$. It is easy to see that each element of such a form belongs to the module $V + W$, where W is the $K_\mathscr{D}$-submodule of $\mathscr{D}_*(i, j)$ generated by all elements of the form $b \otimes u$, where $b \in \mathfrak{M}_k \setminus \{1_k\}$ for certain $k \notin I$ and u is an arbitrary element of $\mathscr{D}(i, j)$. Moreover, it is clear that the submodules W and $V + [1 \otimes \mathscr{D}(i, j)]$ intersect at zero. It follows that $P_\mathscr{D}(i, j) \cap [1 \otimes \mathscr{D}(i, j)] \subseteq (V + W) \cap [1 \otimes \mathscr{D}(i, j)] = V \cap [1 \otimes \mathscr{D}(i, j)]$.

Take an arbitrary element $z \in V \cap [1 \otimes \mathscr{D}(i, j)]$. Without loss of generality, let it have the form (4). Then for all elements x_I of $\mathscr{D}_I$ different from 1_I, the equality

$$\sum_{k \in I} D_{1_k} f_{(k, x_I)} = 0$$

holds.

By (5) and condition (∗∗) we obtain that the last equality is valid for $x_I = 1_I$ as well. Therefore, $z = 0$. Theorem 1 is proved.

In connection with the example of a nonrepresentable triangular category given above, we observe that condition (3) is equivalent to condition (∗∗) in the case when the interval $\langle i, j \rangle$ consists of two objects k and l, while condition (2) is equivalent to condition (∗∗) in the case when the interval $\langle i, j \rangle$ is a singleton.

§3. Representations of graded algebras

In this section we study graded T_n-algebras (with identity), that is, graded algebras (with identity) of the form

$$R = \bigoplus_{0 \leq i < n} R_i, \tag{6}$$

embeddable with preservation of grading (and identity) into the algebra of upper triangular $(n \times n)$-matrices with a natural diagonal grading over a commutative algebra (with identity). An identity of a graded algebra is understood as an identity of the algebra that belongs to its zeroth homogeneous component. It is easy to understand that "adjoining identity" makes it possible to restrict ourselves only to the study of graded T_n-algebras with identity. Thus, till the end of this section, let all algebras have identity and all homomorphisms of algebras preserve the identity and grading (if there is one). A graded algebra of the form (6) is denoted by R.

A right (left) translation of a graded algebra R is defined as any endomorphism θ of the algebra R_0 such that

$$ax = x\theta^k(a) \quad (\text{respectively}, \ xa = \theta^k(a)x)$$

for any $a \in R_0$, $x \in R_k$, $0 \leq k < n$. Observe that the existence of a translation implies the commutativity of the algebra R_0.

REMARK 2. Define an endomorphism θ on the zeroth homogeneous component of the graded algebra $T_n(K)$, where K is a commutative algebra, by

$$\theta(\text{diag}\{d_1, \ldots, d_n\}) = \text{diag}\{d_n, d_1, d_2, \ldots, d_{n-1}\}.$$

Obviously, θ is a right translation (and θ^{-1} a left translation) of the graded algebra $T_n(K)$.

The order of a small category is defined as the cardinality of the set of its objects. The objects of a triangular category of order n will be denoted by numbers $1, 2, \ldots, n$ taken with the natural order.

REMARK 3. Clearly, a Φ-category of order n is representable if and only if the matrix algebra over it is embeddable with preservation of components in the algebra of matrices of order n over a commutative algebra (preservation of components means that a matrix with a zero (i, j)-component becomes a matrix with a zero (i, j)-component). As we show further (see Proposition 3), for the representability of a triangular category of order n it suffices that the matrix algebra over it (with the natural diagonal grading) be a graded T_n-algebra.

REMARK 4. For a graded algebra R such that R_0 is a commutative algebra we define a triangular category $\mathscr{D}^R$ of order n. As a module $\mathscr{D}^R(i, j)$ of morphisms from an object i into j $(i \leq j)$ we take an isomorphic copy of the homogeneous component R_{j-i}; here the composition of morphisms is induced by the multiplication in the algebra R. The mapping that transforms the element $\sum_i r_i$ $(r_i \in R_i)$ into a triangular matrix $(d_{ij})_{i \leq j}$, where d_{ij} is the image of the element $r_{j-i} \in R_{j-i}$ in $\mathscr{D}^R(i, j)$ determines an embedding of the graded algebras $R \to M(\mathscr{D}^R)$.

If moreover, R is an algebra with right (left) translation, then the algebra $M(\mathscr{D}^R)$ also has a right (left) translation. Indeed, without loss of generality let θ be a right translation of the algebra R. Then the correspondence

$$\text{diag}\{d_1, \ldots, d_n\} \mapsto \text{diag}\{\theta(d_n), \theta(d_1), \ldots, \theta(d_{n-1})\}$$

induces a right translation in the algebra $M(\mathscr{D}^R)$.

REMARK 5. Considering homogeneous components of the algebra R as left modules over R_0, we induce on each module $\mathscr{S}(i,j)$ of morphisms from an object i into j of the category $\mathscr{S} = \mathscr{D}^R$ the structure of a left $K_{\mathscr{S}}$-module. This easily implies the injectivity of the module homomorphism $\mathscr{S}(i,j) \to K_{\mathscr{S}} \otimes \mathscr{S}(i,j)$ $(d \mapsto 1 \otimes d)$. It follows that the canonical functor from the category $\mathscr{S}$ into its $K_{\mathscr{S}}$-hull is univalent.

PROPOSITION 3. *The following conditions are equivalent for a triangular category $\mathscr{D}$ of order n:*
1) *the category $\mathscr{D}$ is representable;*
2) $M(\mathscr{D})$ *is a graded T_n-algebra;*
3) *the graded algebra $M(\mathscr{D})$ is embeddable in a graded algebra with translation.*

PROOF. The implications 1) $\Rightarrow$ 2) and 2) $\Rightarrow$ 3) follow immediately from Remarks 3 and 2, respectively. We prove the implication 3) $\Rightarrow$ 1). Without loss of generality, let the algebra $M(\mathscr{D})$ be embeddable in a graded algebra R with a left translation θ. Then the triangular category $\mathscr{D}$ is naturally embeddable in the category $\mathscr{D}^R$. Clearly, $\mathscr{D}^R$ is a triangular category with left translation; here the $(i, i-k)$-translation is induced by the endomorphism θ^k. It remains to use Lemma 1 and Remark 5. Proposition 3 is proved.

COROLLARY 2. *The following conditions are equivalent for a graded algebra R of the form (6):*
1° *the algebra R is a graded T_n-algebra;*
2° *the algebra R is embeddable in a graded algebra with translation;*
3° *the category $\mathscr{D}^R$ is representable;*
4° *the algebra R is embeddable in a graded algebra $A = \bigoplus_i A_i$ $(0 \le i < n)$ such that A_0 is a commutative algebra, $A_i = A_1^i$ $(0 < i < n)$, and for all $1 \le k, m < n$, $a \in A_0$, $x_t \in A_k$, $y_t \in A_m$ the following implication holds:*

$$\sum_t x_t y_t = 0 \Rightarrow \sum_t x_t a y_t = 0. \qquad (7)$$

PROOF. The implication 1° $\Rightarrow$ 2° follows from Remark 2.

2° $\Rightarrow$ 3°. An embedding $R \hookrightarrow H$ of graded algebras naturally induces an embedding $M(\mathscr{D}^R) \hookrightarrow M(\mathscr{D}^H)$ of graded algebras.

Besides, by Remark 4, if H is an algebra with translation, then $M(\mathscr{D}^H)$ is an algebra with translation as well. It remains to use the implication 3) $\Rightarrow$ 1) from the previous proposition.

3° $\Rightarrow$ 1°. Indeed, by Remark 4, the algebra R can be embedded in the graded algebra $M(\mathscr{D}^R)$. Now the desired statement follows from Remark 3.

1° $\Rightarrow$ 4°. Direct verification shows that one can take $T_n(K)$ for the algebra A. Here the validity of quasi-identities of the form (7) is easily verified by means of the translation defined in Remark 2.

4° $\Rightarrow$ 3°. It is easy to notice that the category $\mathscr{D}^A$ satisfies all conditions of Proposition 2. Corollary 2 is proved.

For an algebra R let $R_{0,t}$ denote the set of strings of the form $(x_1, \ldots, x_t)$ $(x_i \in R_0)$, and let $(x)_t = (x_1, \ldots, x_t)$, $(1)_t = (1, \ldots, 1)$. For a finite subset $f \subseteq R_k \times R_l$ and an element $a \in R_0$ let $D_a f$ denote the sum of the elements uav, where the pair (u, v) runs over the set f.

COROLLARY 3. *Let R be a graded algebra of the form (6) such that R_0 is a free module that possesses a base containing the identity of the algebra R. Then R is a graded T_n-algebra if and only if R_0 is a commutative algebra and for every $t \in \{1, \ldots, n-2\}$ the following condition holds in the algebra R:*

(∗∗∗) *Let, for all finite subsets $f(k, (x)_t) \subseteq R_k \times R_{t-k+1}$ ($k \in \{1, \ldots, t\}$, $(x)_t \in R_{0,t}$), only finitely many among which are nonempty, all equalities of the form*

$$\sum_{k=1}^{t} D_1 f(k, (x)_t) = 0, \qquad \sum_{y_r \in R_0} D_{y_r} f(r, (y)_t) = 0$$

(where $(x)_t, (y)_t \in R_{0,t}$, $(x)_t \neq (1)_t$, $r \in \{1, \ldots, t\}$) hold. Then the equality

$$\sum_{k=1}^{t} D_1 f(k, (1)_t) = 0$$

holds in the algebra R.

PROOF. Indeed, the requirement of the commutativity of R_0 raises no doubts. Besides, simple verification shows that the triangular category $\mathscr{D}^R$ satisfies condition (∗∗) of Theorem 1 for objects i, j ($i < j$) if and only if the algebra R satisfies condition (∗∗∗) for $t = j - i - 1$. It remains to apply Theorem 1 and the equivalence $1° \Leftrightarrow 3°$ of the previous Corollary. Corollary 3 is proved.

Chapter 2
Representation of Algebras by Triangular Matrices. Algebras with Diagonal

In this chapter necessary and sufficient conditions are found for the existence of an embedding $\varphi: R \to T_n(K)$ (K is a commutative algebra) such that $\ker \varphi_{ii} = I_i$ ($i = 1, \ldots, n$) are fixed ideals of the algebra R that contain its commutator ideal (see Theorem 2). As a corollary we obtain a criterion for T_n^M-representability of the algebra R ($M \subseteq \{1, \ldots, n\}$). When $M = \{1, \ldots, n\}$, this criterion yields a positive answer to A. Z. Anan'in's hypothesis (see Corollary 4).

In addition, we study representations of algebras with diagonal that satisfy an identity with diagonal.

§1. Statement of the basic theorem and its corollaries

Let, until the end of this chapter, $X = X_1 \cup \cdots \cup X_n$ be the union of pairwise disjoint countable sets of free variables; let $R\langle X\rangle$ be the algebra of generalized polynomials over an algebra R with variables from X (we assume that $X \subseteq R\langle X\rangle$), and let $\mathfrak{X}$ (respectively, $\mathfrak{X}^{\cdot}$) be the commutative semigroup with identity (without identity) freely generated by the set X. A partial derivative of a generalized polynomial f with respect to variables $x, y, \ldots, z$ will be denoted by $\partial f/\partial w$, where $w = xy \cdots z$ (we assume that $\partial f/\partial 1 = f$). For a generalized polynomial f let X^f denote the set of variables $x \in X$ on which f depends essentially, that is, $f \notin R\langle X\setminus\{x\}\rangle$. Let $X_i^f = X^f \cap X_i$. A (partial) mapping ξ from X into R is called a (partial) interpretation; let $f\xi$ denote the result of replacing each variable $x \in \operatorname{dom} \xi$ in a generalized polynomial f by $\xi(x)$.

A generalized polynomial from $R\langle X\rangle$ is called an n-polynomial over R if it is decomposable into a sum of monomials of the form $a_1 x_1 \cdots a_n x_n a_{n+1}$ such that $a_i \in R^{\#}$, $x_j \in X_j$. In particular, an n-polynomial over Φ is a linear combination of monomials of the form $x_1 x_2 \cdots x_n$ ($x_j \in X_j$). In the sequel, unless otherwise stated, an n-polynomial means an n-polynomial over R.

Note that if $\partial f/\partial w \neq 0$, where f is an n-polynomial and $w \in \mathfrak{X}$, then the monomial w is of the form $x_{i_1} \cdots x_{i_k}$, where $i_1 < \cdots < i_k$, $x_{i_\nu} \in X_{i_\nu}^f$.

A homomorphism $\varphi: R \to T_n(K)$ $(r \mapsto (\varphi_{ij}(r)))$ of algebras is called *compatible* with a sequence $I_1, \ldots, I_n$ of ideals of the algebra R if $\ker \varphi_{ii} \supseteq I_i$ for $i = 1, \ldots, n$; it is called *strictly compatible* if moreover, the opposite inclusions also hold.

The basic result of this chapter is the following

THEOREM 2. *Let R be an algebra over a commutative regular ring, let $I = (I_1, \ldots, I_n)$ be a sequence of its ideals, each of which contains the commutator ideal of the entire algebra. Then the following conditions are equivalent*:

a) *there exists a strictly compatible with the sequence I embedding of the algebra R into the algebra of upper triangular matrices of order n over a commutative algebra*;

b) *for every n-polynomial f over R and an arbitrary interpretation $\xi: X \to R$ the following implication holds*:

$$\bigwedge_w ((\partial f/\partial w)\xi = 0) \Rightarrow f\xi = 0, \qquad (8)$$

where w runs over all nonempty monomials from $\mathfrak{X}(f, \xi, I) := \{x_{i_1} \cdots x_{i_k} \in \mathfrak{X} \mid i_1 < \cdots < i_k, \ k \geq 0, \ x_{i_\nu} \in X_{i_\nu}^f, \ \xi(x_{i_\nu}) \notin I_{i_\nu}\}$;

c) *for every n-polynomial f over Φ and an arbitrary interpretation $\xi: X \to R$ the implication (8) holds.*

Clearly, if condition b) holds for a sequence $(I_1, \ldots, I_n)$ of ideals, then this condition holds for every sequence $(J_1, \ldots, J_n)$ of ideals of the algebra R such that $J_1 \subseteq I_1, \ldots, J_n \subseteq I_n$. Therefore, it follows from Theorem 2 that condition (a) is equivalent to a condition obtained by dropping the word "strictly" from it.

Note that the implication b) $\Rightarrow$ c) is trivial.

The particular cases of Theorem 2, when each of the ideals in the sequence I coincides either with R or with its commutator ideal, are of principal interest.

In this connection, a generalized polynomial is called an (n, M)-polynomial $(M \subseteq \{1, \ldots, n\})$, if it is obtained from an n-polynomial g by replacing in it the variables from the sets X_i^g $(i \notin M)$ by suitable elements of R.

THEOREM 3. *An algebra R over a commutative regular ring is a T_n^M-algebra if and only if for every (n, M)-polynomial g and an arbitrary interpretation $\eta: X \to R$, the following condition holds*:

$$\bigwedge_{w \in \mathfrak{X}^*} \left(\frac{\partial g}{\partial w}\eta = 0\right) \Rightarrow g\eta = 0. \qquad (9)$$

PROOF. *Necessity.* Let $\varphi: R \to T_n^M(K)$ be an embedding of algebras, where K is a commutative algebra. In Theorem 2 assume that $I_i = \ker \varphi_{ii}$. Since $\varphi_{ii}: R \to K$ is a homomorphism of algebras, we obtain $(R, R) \subseteq I_i$. Now the desired statement follows straightforwardly from the implication a) $\Rightarrow$ b) (we need only take into consideration the fact that $I_i = R$ for $i \notin M$).

Sufficiency. Let $I_i = (R, R)$ for $i \in M$ and $I_i = R$ otherwise in Theorem 2. By Theorem 2, we need only verify condition b). Clearly, in our case condition b) is equivalent to the condition: for every (n, M)-polynomial f and an arbitrary interpretation $\xi: X \to R$ implication (8) holds, in which the conjunction is taken over all monomials $w \in \mathcal{X}^*$ that do not contain variables from the set $X_\xi^f := \{x \in X^f | \xi(x) \in (R, R)\}$.

Let the antecedents of this implication hold. For each variable $x \in X_\xi^f$ choose elements $a_{x,i} \in R^\#$, $b_{x,i}, c_{x,i} \in R$ $(1 \leq i \leq p)$ such that

$$\xi(x) = \sum_{i=1}^{p} a_{x,i}[b_{x,i}, c_{x,i}].$$

Let g denote the (n, M)-polynomial obtained from f as a result of replacing each variable $x \in X_\xi^f$ by the polynomial

$$\sum_{i=1}^{p} a_{x,i}[y_{x,i}, c_{x,i}],$$

where $y_{x,i}$ $(x \in X_\xi^f, 1 \leq i \leq p)$ are different variables from $X \setminus X^f$ such that $y_{x,i} \in X_j$ for $x \in X_j$. Define an interpretation $\eta: X \to R$ that differs from the interpretation ξ only maybe on the set $Y = \{y_{x,i} | x \in X_\xi^f, 1 \leq i \leq p\}$ by setting $\eta(y_{x,i}) = b_{x,i}$.

Observe that $(\partial g / \partial w)\eta = 0$ for all $w \in \mathcal{X}^*$. Indeed, if a monomial $w \in \mathcal{X}^*$, does not contain variables from $Z = Y \cup X_\xi^f$, then we obtain

$$\frac{\partial g}{\partial w}\eta = \left(\frac{\partial g}{\partial w}\eta'\right)\eta'' = \frac{\partial(g\eta')}{\partial w}\eta'' = \frac{\partial(f\xi')}{\partial w}\xi'' = \frac{\partial f}{\partial w}\xi = 0$$

where η' and ξ' (η'' and ξ'') denote restrictions of the interpretations η and ξ to the set Z (the set $X \setminus Z$), respectively. Here we have used the equalities $g\eta' = f\xi'$, $\eta'' = \xi$ and the permutability of a partial interpretation and differentiation with respect to variables that do not belong to the domain of this partial interpretation.

If a monomial w contains a variable from Z, then $\partial g / \partial w = 0$. Indeed, since every variable from Z that essentially appears in the polynomial g belongs to the set Y, it suffices to observe that $\partial g / \partial y_{x,i} = 0$ for $y_{x,i} \in Y$. This follows directly from the equality $\partial [y_{x,i}, c_{x,i}] / \partial y_{x,i} = 0$ and the fact that g is decomposable into a sum of a generalized polynomial that does not contain the given variable $y_{x,i}$ and polynomials of the form $u[y_{x,i}, c_{x,i}]v$, where u and v are generalized monomials, possibly absent, which do not contain the variable $y_{x,i}$.

Now the consequent of the implication we are verifying follows from the implication (9) and the equality $f\xi = g\eta$. Theorem 3 is proved.

If $M = \{1, \ldots, n\}$, Theorem 3 yields a positive answer to A. Z. Anan'in's conjecture.

COROLLARY 4. *An algebra R over a commutative regular ring is a T_n-algebra if and only if for every n-polynomial g over R and an arbitrary interpretation $\eta: X \to R$ implication (9) holds.*

As an illustration, we show how this statement implies the following

THEOREM (Anan'in [1]). *An algebra R (over a field) that satisfies the identities*

$$[x_1, \ldots, x_t]y_1 \cdots y_t[z_1, \ldots, z_t] = 0, \qquad [x_1, y_1] \cdots [x_t, y_t] = 0, \qquad (10)$$

is embeddable in the algebra of upper triangular matrices over a commutative algebra.

The fact that a positive answer to Anan'in's conjecture implies this theorem was known to A. Z. Anan'in.

Let $R^{(0)} = R^{\#}$, and let $R^{(k+1)}$ be the ideal of the algebra R generated by all elements of the form $[a, b]$ ($a \in R^{(k)}$, $b \in R$). It is easy to observe that if the algebra R satisfies the identities (10), then it satisfies the equality $R^{(k)}R^m R^{(l)} = 0$ for certain numbers k, m, and l depending on t only. Therefore, to prove the preceding theorem, it suffices to prove, by Corollary 4, that if f is an n-polynomial over R and $\xi: X \to R$ is an interpretation of the variables such that $(\partial f/\partial w)\xi = 0$ for all $w \in \mathfrak{X}^*$, then $f\xi \in R^{(k)}R^m R^{(l)}$ ($n = k + m + l$).

To this end, we define linear operators L and Ψ on the algebra $R\langle X \rangle$, which are identical on the subalgebra R, by setting

$$L(axg) = [a, x]g, \qquad \Psi(gxa) = g(x, a),$$

where $a \in R^{\#}$, $x \in X$, $g \in R^{\#}\langle X \rangle$.

Let $\mathscr{A}(i, j)$ denote the subspace generated by elements of the form $a_i x_{i+1} a_{i+1} \cdots x_{n-j} a_{n-j}$ ($x_k \in X_k^f$, $a_{i+1}, \ldots, a_{n-j-1} \in R^{\#}$, $a_i \in R^{(i)}$, $a_{n-j} \in R^{(j)}$). Let ξ_i denote the restriction of the interpretation ξ to the set X_i^f.

Let $f_0 = f$, $f_i = (L(f_{i-1}))\xi_i$ ($i \leq n$). Clearly, $(L(\mathscr{A}(i-1, 0))\xi_i \subseteq \mathscr{A}(i, 0)$. Therefore, $f_i \in \mathscr{A}(i, 0)$. Using induction on i, we show that $(\partial f_i/\partial w)\xi = 0$ for $w \in \mathfrak{X}^*$ and $f_i\xi = f\xi$. This statement is trivial for $i = 0$. Now let $i > 0$. If a monomial $w \in \mathfrak{X}$ contains at least one variable from the set X_i^f, then $\partial f_i/\partial w = 0$ (since $f_i \in \mathscr{A}(i, 0)$).

Now if a monomial $w \in \mathfrak{X}$ does not depend on variables from X_i^f, then the operator $\partial/\partial w$ is permutable with the partial interpretation ξ_i. Also, since $f_{i-1} \in \mathscr{A}(i-1, 0)$,

$$L(f_{i-1}) = f_{i-1} - \sum_{x \in X_i^f} x \frac{\partial f_{i-1}}{\partial x}.$$

Applying the partial interpretation ξ_i and the operator $\partial/\partial w$ (and taking into consideration their permutability) to this equality we obtain

$$\frac{\partial f_i}{\partial w} = \frac{\partial f_{i-1}}{\partial w} \xi_i - \sum_{x \in X_i^f} \left(x \frac{\partial f_{i-1}}{\partial (xw)} \right) \xi_i.$$

Using the induction hypothesis, from this we obtain that $(\partial f_i/\partial w)\xi = 0$ for $w \in \mathfrak{X}^*$ and $f_i \xi = f_{i-1} \xi = f \xi$.

Now let $f_{k,0} = f_k$, $f_{k,i+1} = (\Psi(f_{k,i}))\xi_{n-i}$ ($k < n-i$). Analogously we verify that $f_{k,i} \in \mathscr{A}(k, i)$, $(\partial f_{k,i}/\partial w)\xi = 0$ for $w \in \mathfrak{X}^*$ and $f_{k,i}\xi = f\xi$. In particular, $f\xi = f_{k,l}\xi \in (\mathscr{A}(k, l))\xi \subseteq R^{(k)} R^m R^{(l)}$. Q.E.D.

§2. Algebras with diagonal, adjoining a diagonal

LEMMA 2. *Let $a^{(1)}, \ldots, a^{(n)} \in T_n(A)$, A an arbitrary ring, and $a_{ii}^{(i)} = 0$ ($i = 1, \ldots, n$). Then $a^{(1)} \cdots a^{(n)} = 0$.*

PROOF. It is easy to prove by induction on n that for every nondecreasing sequence $i_0, \ldots, i_n$ ($1 \le i_0$, $i_n \le n$) of integers there exists a number t ($1 \le t \le n$) such that $i_{t-1} = i_t = t$. Now $a_{i_0 i_1}^{(1)} a_{i_1 i_2}^{(2)} \cdots a_{i_{n-1} i_n}^{(n)} = 0$, whence we have the desired equality. Lemma 2 is proved.

It follows from this lemma that in the algebra $T_n(K)$, where K is a commutative algebra, the equality

$$(x_1 - d_1(x_1)) \cdots (x_n - d_n(x_n)) = 0 \tag{11}$$

holds, where d_k is a homomorphism from the algebra $T_n(K)$ into its center that transfers a matrix $x = (x_{ij})$ into a scalar matrix $x_{kk} E_n$. It is clear that

$$d_i d_j = d_j. \tag{12}$$

That gives rise to the following definitions. A diagonal mapping of an algebra R is defined as any homomorphism from the algebra R into its center $Z(R)$. A sequence $d = (d_1, \ldots, d_n)$ of diagonal mappings of the algebra R that satisfy equalities (12) is called a diagonal of order n. An algebra with a fixed diagonal d of order n is denoted by $\langle R, d \rangle_n$. A homomorphism $\varphi: \langle R, D \rangle_n \to \langle S, d \rangle_n$ of algebras with diagonal is understood as a homomorphism $\varphi: R \to S$ of algebras that preserves the diagonal mappings, that is, $\varphi D_i = d_i \varphi$. Thus, a category of algebras with diagonal of order n is defined.

An identity of an algebra $\langle R, d \rangle_n$ is understood as an identity of the algebra R that is preserved under the homomorphisms d_i.

The equality (11) is called an identity with diagonal of order n or an identity with diagonal d.

Let $d = (d_1, \ldots, d_n)$ be a sequence of diagonal mappings of an algebra R, $\xi: X \to R$ an interpretation, and $w = x_{i_1} \cdots x_{i_k} \in \mathfrak{X}$ ($x_{i_\nu} \in X_{i_\nu}$). Then set $d(w)\xi = d_{i_1}(x_{i_1}\xi) \cdots d_{i_k}(x_{i_k}\xi)$, $d(1)\xi = 1$.

LEMMA 3. *Let $d = (d_1, \ldots, d_n)$ be a sequence of diagonal mappings of an algebra R in which the identity (11) is satisfied. Then condition* b) *of Theorem 2 holds for the algebra R and the sequence $I = (\ker d_1, \ldots, \ker d_n)$ of its ideals.*

PROOF. It follows from the identity (11) in the algebra R and the equalities

$$xy - d_k(xy) = d_k(y)(x - d_k(x)) + x(y - d_k(y))$$

that for every $a_0, \ldots, a_n \in R^\#$, $b_1, \ldots, b_n \in R$ the equality

$$a_0(b_1 - d_1(b_1))a_1 \cdots (b_n - d_n(b_n))a_n = 0$$

holds.

Consider an n-monomial $g = a_0 x_1 a_1 \cdots x_n a_n \in R\langle X \rangle$. Removing the parentheses in the previous equality shows that, for every interpretation $\xi: X \to R$, the equality

$$\sum_{\substack{1 \le i_1 < \cdots < i_k \le n \\ k \ge 0}} (-1)^k \frac{\partial^k g}{\partial x_{i_1} \cdots \partial x_{i_k}} \xi \, d(x_{i_1} \cdots x_{i_k})\xi = 0$$

holds, in which $b_i = \xi(x_i)$.

Therefore, for an arbitrary n-polynomial f, the sum

$$\sum_{w \in \mathfrak{X}} (-1)^{|w|} \frac{\partial f}{\partial w} \xi \, d(w)\xi$$

(where $|w|$ denotes the length of the monomial w) equals zero. If the antecedent of the implication (8) holds now (here $I_i = \ker d_i$), then all summands with $w \ne 1$ in the last sum equal zero, and hence, $f\xi = 0$. Lemma 3 is proved.

Until the end of this section, let R be an algebra, and let $I = (I_1, \ldots, I_n)$ be a sequence of its ideals, each of which contains the commutator ideal of the algebra R. A homomorphism φ from the algebra R into the algebra $\langle S, D \rangle_n$ is said to be compatible with the sequence I if $D_i \varphi(I_i) = 0$ for $i = 1, \ldots, n$.

Define an algebra $\langle R_I, d\rangle_n$ with identity, taking for R_I the tensor product $R^\# \otimes (R^\#/I_1) \otimes \cdots \otimes (R^\#/I_n)$ of algebras (here the algebra R is identified with the subalgebra $R \otimes 1 \otimes \cdots \otimes 1 \subseteq R_I$ and we assume that

$$d_i(r_0 \otimes (r_1 + I_1) \otimes \cdots \otimes (r_n + I_n))$$
$$= 1 \otimes (r_1 + I_1) \otimes \cdots \otimes (r_0 r_i + I_i) \otimes \cdots \otimes (r_n + I_n) \ (r_i \in R^\#).$$

Put $K_I = \operatorname{im} d_1$; it is easily seen that d_i $(i = 1, \ldots, n)$ are epimorphisms of K-algebras $R_I \to K_I$.

Clearly, the embedding $\varepsilon_{R,I}: R \to R_I$ $(r \mapsto r \otimes 1 \otimes \cdots \otimes 1)$ is universal in the sense that, for every homomorphism φ of R into an algebra $\langle S, D\rangle_n$ with identity compatible with the sequence I, there exists a uniquely determined homomorphism $\psi: \langle R_I, d\rangle_n \to \langle S, D\rangle_n$ of algebras with diagonal preserving the identity such that $\varphi = \psi \varepsilon_{R,I}$.

All elements of the form $(x_1 - d_1(x_1))(x_2 - d_2(x_2)) \cdots (x_n - d_n(x_n))$ $(x_i \in R_I)$ generate an ideal $\mathscr{F}_{R,I}$ in the algebra R_I which, in the case when $I_1 = \cdots = I_n = (R, R)$, is denoted by $\mathscr{F}_R$.

Since the diagonal mappings $d_1, \ldots, d_n$ of the algebra R_I annihilate the ideal $\mathscr{F}_{R,I}$, a structure of algebra with diagonal

$$\overline{d} \quad (\overline{d}_i(x + \mathscr{F}_{R,I}) = d_i(x) + \mathscr{F}_{R,I})$$

satisfying identity (11) is induced on the factor algebra $R_I/\mathscr{F}_{R,I}$. It follows from the universality of the homomorphism $\varepsilon_{R,I}$ that the existence of an embedding compatible with the sequence I from the algebra R into an algebra with diagonal of order n that satisfies identity (11) is equivalent to the equality

$$R \cap \mathscr{F}_{R,I} = 0. \tag{13}$$

We give a characterization of the elements from the left-hand side of this equality. To this end we need the following

LEMMA 4. *The kernel of the natural epimorphism of the modules*

$$A_1 \otimes \cdots \otimes A_r \to A_1/B_1 \otimes \cdots \otimes A_r/B_r,$$
$$a_1 \otimes \cdots \otimes a_r \mapsto (a_1 + B_1) \otimes \cdots \otimes (a_r + B_r)$$

(*where B_i is a submodule of the module A_i $(i = 1, \ldots, r)$) is generated (as a module) by all elements of the form $a_1 \otimes \cdots \otimes a_r$, where $a_1 \in A_1, \ldots, a_r \in A_r$, and $a_i \in B_i$ for certain $i \in \{1, \ldots, r\}$.*

The proof of this lemma is completely analogous to the proof of the fact that the functor of the tensor product is right exact.

Let Δ_i denote the endomorphism of the K_I-module R_I that transforms an element a into $a - d_i(a)$, and let Δ be the module homomorphism $R^{\otimes n} \to R_I$ $(a_1 \otimes \cdots \otimes a_n \mapsto \Delta_1(a_1) \cdots \Delta_n(a_n))$. Extend the correspondence

$$a_1 \otimes \cdots \otimes a_n \mapsto \frac{\partial^k (x_1 \cdots x_n)}{\partial x_{i_1} \cdots \partial x_{i_k}} \xi \, d(x_{k_1} \cdots x_{i_k}) \xi$$

(where $i \leq i_1 < \cdots < i_k \leq n$, $x_i \in X_i$, $\xi(x_i) = a_i \in R$) to a module homomorphism $d_M \colon R^{\otimes n} \to R_I$ ($M = \{i_1, \ldots, i_k\}$).

PROPOSITION 4. $1°$. $R \cap \mathscr{F}_{R,I} = d_\varnothing(\bigcap_{\varnothing \neq M \subseteq \{1,\ldots,n\}} \ker d_M)$.

$2°$. *Condition* b) *of Theorem* 2 *follows from equality* (13).

$3°$. *In the case of algebras over a field, equality* (13) *follows from condition* (c) *of Theorem* 2.

PROOF. $1°$. The inclusion $\supseteq$ is obvious since
$$\Delta = \sum_{M \subseteq \{1,\ldots,n\}} (-1)^{|M|} d_M.$$

We check the opposite inclusion. Using the obvious identities
$$\Delta_k(xy) = x\Delta_k(y) + \Delta_k(x)d_k(y), \qquad [x, \Delta_k(y)] = \Delta_k([x,y]), \qquad (14)$$
and the fact that an arbitrary element of R_I is decomposable into a sum of elements of the form $r_0 d_1(r_1) \cdots d_n(r_n)$ ($r_i \in R^\#$), we easily establish that $\mathscr{F}_{R,I} = R^\# \Delta_1(R) \cdots \Delta_n(R)$. Now let
$$r = \sum_{\mathfrak{M}} a_0 \Delta_1(a_1) \cdots \Delta_n(a_n) \in R,$$
where the summation is taken over the finite set $\mathfrak{M}$ of sequences $(a_0, \ldots, a_n)$ such that $a_0 \in R^\#$, $a_1, \ldots, a_n \in R$. Then
$$\sum_{\mathfrak{M}} a_0 d_1(a_1) \cdots d_n(a_n) = 0.$$

By the previous lemma, we can assume without loss of generality that $\mathfrak{M}$ consists of sequences $(a_0, \ldots, a_n)$ such that $a_i \in I_i$ for certain $i \in \{1, \ldots, n\}$. Taking into consideration the second identity in (14) as well as the equality $c\Delta_i(a) = \Delta_i(ca)$ ($a \in I_i$, $c \in R^\#$), we obtain that $r = \Delta(b)$ for certain $b \in R^{\otimes n}$. Since $\Delta(b) \in R$, it follows that $d_M(b) = 0$ for every nonempty set $M \subseteq \{1, \ldots, n\}$, and hence, $\Delta(b) = d_\varnothing(b)$. Therefore, the element r belongs to the right-hand side of the equality we are verifying.

$2°$. By the previous lemma, condition b) holds in the algebra $\langle R_I/\mathscr{F}_{R,I}, \overline{d}\rangle_n$ for the sequence of ideals $(\ker \overline{d}_1, \ldots, \ker \overline{d}_n)$. Now if equality (13) holds, we may identify the algebra R with its image under the superposition of the homomorphisms $R \to R_I \to R_I/\mathscr{F}_{R,I}$. It remains to apply the fact that condition b) carries over to the subalgebras and that $\{r \in R \mid d_i(r) \in \mathscr{F}_{R,I}\} = \{r \in R \mid d_i(r) = 0\} = I_i$.

$3°$. Let the ground ring be a field. Choose a basis $\mathfrak{M}_i$ of the algebra R modulo the ideal I_i. By part $1°$ of this proposition, an arbitrary element λ of $R \cap \mathscr{F}_{R,I}$ can be represented in the form
$$\lambda = \sum_{t=1}^m \alpha_t \Delta(b_{1t} \otimes \cdots \otimes b_{nt}) \qquad (\alpha_t \in \Phi, \ b_{it} \in \mathfrak{M}_i \cup I_i).$$

Consider n-monomials $g_t = \alpha_t x_{1t} \cdots x_{nt}$ $(t = 1, \ldots, m)$ over Φ, where $x_{it} \in X_i$ and the variables x_{it} and $x_{it'}$ coincide if and only if $b_{it} = b_{it'}$. Choose an interpretation $\xi: X \to R$ such that $\xi(x_{it}) = b_{it}$ $(i = 1, \ldots, n; \ t = 1, \ldots, m)$.

Clearly,
$$\lambda = \sum_{w \in \mathfrak{X}(f,\xi,I)} (-1)^{|w|} \frac{\partial f}{\partial w} \xi d(w) \xi, \tag{15}$$

where $f = g_1 + \cdots + g_m$ is an n-polynomial over Φ (recall that $\mathfrak{X}(f, \xi, I) = \{x_{i_1} \cdots x_{i_k} \in \mathfrak{X} \mid i_1 < \cdots < i_k, \ k \geq 0, \ x_i \in X_i^r, \xi(x_i) \notin I_i\}$).

Observe that, by the inclusion $\xi(X_i^f) \subseteq \mathfrak{M}_i \cup I_i$, and the injectivity of the mapping $\xi|_{X_i^f}$, the elements of the form $d(w)\xi$ $(w \in \mathfrak{X}(f, \xi, I))$ are linearly independent over Φ, and hence also over $R^{\#}$, in the algebra R_I. Since $\lambda \in R$, equality (15) implies $\lambda = f\xi$ and $(\partial f / \partial w)\xi = 0$ for $w \in \mathfrak{X}(f, \xi, I) \setminus \{1\}$. Now condition c) of Theorem 2 produces the equality $f\xi = 0$, that is, $\lambda = 0$. Proposition 4 is proved.

In conclusion of this section we observe that, in the case of algebras over an arbitrary commutative ring with identity, we can translate the equality $R \cap \mathscr{F}_R = 0$ into the language of quasi-identities of the algebra R using Lemma 4 and part 1° of Proposition 4. However, the quasi-identities we would obtain would have a very inconvenient form and, in general, would contain constants from the ground ring.

§3. Proof of the basic theorem

In this section R is an algebra with a sequence $I = (I_1, \ldots, I_n)$ of ideals, each of which contains the commutator ideal of the algebra R.

Considering the algebra of upper triangular $(n \times n)$-matrices over a commutative algebra as an algebra with natural diagonal that satisfies, by Lemma 2, identity (11), we easily convince ourselves that the implication a) $\Rightarrow$ b) follows from the universality of the homomorphism $\varepsilon_{R,I}$ and part 2° of Proposition 4.

The implication b) $\Rightarrow$ c) is trivial. Therefore, it remains to prove the implication c) $\Rightarrow$ a). First, we consider the case of algebras over a field Φ. Extend the diagonal mappings d_i $(i = 1, \ldots, n)$ of the algebra $\mathscr{R}_0 = R_I$ to diagonal mappings of the algebra of polynomials $\mathscr{R} = \mathscr{R}_0[q, q_1, \ldots, q_n]$ (of commutative variables $q, q_1, \ldots, q_n$) by setting $d_i(q) = q_i$, $d_i(q_j) = q_j$. Thus a diagonal of order n is defined on the algebra $\mathscr{R}$. By means of that diagonal, we define linear mappings $\Delta_i: \mathscr{R} \to \mathscr{R}$ $(r \mapsto r - d_i(r))$ for $i = 1, \ldots, n$.

Let H denote the subalgebra of $\mathscr{R}$ generated by the subalgebra $R^{\#} \otimes 1 \otimes \cdots \otimes 1$ and a variable q $(H \approx R^{\#}[q])$ and set $V_{ij} = d_i(H) \cdots d_j(H) \subseteq \mathscr{R}$ $(i \leq j)$. For the sake of convenience assume that $V = V_{1n}$ and $V_{ij} = \Phi$ for $i > j$. Clearly, d_i is an epimorphism of V-algebras $\mathscr{R} \to V$ and Δ_i is a V-linear mapping.

Let $Q_0 = H$, $Q_i = H\Delta_1(H)\cdots\Delta_i(H)$ $(i = 1, \ldots, n)$. Using identities of the form (14), we easily establish that Q_i is an ideal of the subalgebra $V_{1i}H \subseteq \mathscr{R}$. Observe that the space $\Delta_i(H)$, and hence, the space Q_i, is contained in the kernel of the homomorphism d_i.

Taking into consideration that $Q_i \subseteq Q_{i-1}d_i(H)$, we choose a direct complement $\mathfrak{N}_i$ to the subspace Q_i in the space $W_i = Q_iV_{i+1n}$, $N_j = \mathfrak{N}_jV_{j+1n}$ $(i = 0, \ldots, n; j = 1, \ldots, n)$. It is clear that W_i is an ideal of the algebra $\mathscr{R}$ and also

$$\mathscr{R} = W_0 \supseteq W_1 \supseteq \cdots \supseteq W_n = Q_n, \qquad d_i(W_i) = 0, \qquad W_{i-1} = N_i \oplus W_i.$$

In particular, $W_{i-1} = N_i \oplus \cdots \oplus N_j \oplus W_j$ $(i \leq j)$. Following the construction analogous to one of the constructions used in [18], for every element $r \in \mathscr{R}$ we take linear mappings $r_{ij}: N_i \to N_j$ $(i \leq j)$ (written to the right of the argument) that are uniquely determined by the congruences

$$hr \equiv hr_{ii} + \cdots + hr_{in} \mod W_n \ (h \in N_i).$$

A simple verification shows that

$$(rs)_{ij} = \sum_{k=i}^{j} r_{ik}s_{kj} \qquad (r, s \in \mathscr{R}). \tag{16}$$

In particular, we have homomorphisms $\psi_i: \mathscr{R} \to \mathrm{End}_\Phi(N_i)$, $r \mapsto r_{ii}$ of Φ-algebras.

LEMMA 5. $\ker \psi_i \cap V_{in} = 0$.

PROOF. Indeed, let $v_{ii} = 0$ for some $v \in V_{in}$. Then $W_{i-1}v \subseteq W_i$ and, in particular, $z = \Delta_1(q)\cdots\Delta_{i-1}(q)v \in W_i$. Since $d_i(W_i) = 0$, we obtain $d_i(z) = (q_i - q_1)(q_i - q_2)\cdots(q_i - q_{i-1})v = 0$. It follows that $v = 0$, since the elements $q_i - q_j$ $(i \neq j)$ are not zero divisors in the algebra $\mathscr{R}$. Lemma 5 is proved.

Now we define subspaces $D(i, j) \subseteq \mathrm{Hom}_\Phi(N_i, N_j)$ $(i \leq j)$ by induction over $j - i$, setting

$$\mathscr{D}(i, i) = \psi_i(V_{in}),$$
$$\mathscr{D}(i, j) = \sum_{r \in R} \mathscr{D}(i, i)r_{ij}\mathscr{D}(j, j) + \sum_{i<k<j} \mathscr{D}(i, k)\mathscr{D}(k, j) \quad (i < j),$$

where the multiplication means superposition of transformations. Clearly, $\mathscr{D}(i, j)\mathscr{D}(j, k) \subseteq \mathscr{D}(i, k)$ for all $i \leq j \leq k$. This determines a triangular category $\mathscr{D}$ over I of order n.

The embedding $V_{i+1n} \subseteq V_{in}$ of algebras induces, by Lemma 5, the embedding $x_{i+1,i}: \mathscr{D}(i+1, i+1) \to \mathscr{D}(i, i)$ $(v_{i+1,i+1} \mapsto v_{ii}, v \in V_{i+1n})$ of algebras.

LEMMA 6. *For all* $\alpha \in \mathscr{D}(i, i+1)$, $\beta \in \mathscr{D}(i+1, i+1)$

$$\alpha\beta = x_{i+1,i}(\beta)\alpha.$$

PROOF. Let $j = i + 1$. Obviously, it suffices to consider the case when $\alpha = r_{ij}$, $\beta = v_{jj}$ ($r \in R$, $v \in V_{jn}$). Since the space $N_i = \mathfrak{N}_i V_{jn}$ sustains multiplication by elements of V_{jn}, we obtain $hv_{ii} = hv$, $(hv)r_{ii} = v(hr_{ii})$ for every element $h \in N_i$. Therefore,

$$hv_{ii}r_{ij} = (hv)r_{ij} \equiv vhr - (vh)r_{ii} = vhr - v(hr_{ii})$$
$$= (hr - hr_{ii})v \equiv (hr_{ij})v \equiv hr_{ij}v_{jj} \mod W_j,$$

whence, $hv_{ii}r_{ij} = hr_{ij}v_{jj}$, $v_{ii}r_{ij} = r_{ij}v_{jj}$, that is, $x_{ji}(\beta)\alpha = \alpha\beta$. Lemma 6 is proved.

Observe that the inclusion $N_i(r - d_i(r)) \subseteq W_i$ ($r \in \mathscr{R}$) implies the equality of mappings $r_{ii} = (d_i(r))_{ii}$. Therefore, $r_{ii} \in \mathscr{D}(i, i)$ for $r \in R$. By (16), the mapping $\varphi: R \to M(\mathscr{D})$, $r \mapsto (r_{ij})_{i \leq j}$, is a homomorphism of algebras and for every $r \in R$

$$r_{ii} = 0 \Leftrightarrow (d_i(r))_{ii} = 0 \Leftrightarrow d_i(r) = 0 \Leftrightarrow r \in I_i \qquad (17)$$

(the equivalence in the middle follows from Lemma 5).

If $r \in R \cap W_n$, then $N_i r \subseteq W_n \subseteq W_j$ for $i \leq j \leq n$; whence, $r_{ij} = 0$ and $\varphi(r) = 0$. Conversely, if $\varphi(r) = 0$, then $r_{in} = 0$; whence $N_i r \subseteq W_n$, and hence $\mathscr{R} r \subseteq W_n$; in particular, $r \in R \cap W_n$. Thus, $\ker \varphi = R \cap W_n = R \cap Q_n$.

It is clear that Q_n is contained in the sum of the subspace $\mathscr{F}_{R,I} \subseteq \mathscr{R}_0$ and the ideal $\mathscr{R}'$ of the algebra $\mathscr{R}$ that is generated by the variables $q, q_1, \ldots, q_n$. Therefore,

$$R \cap Q_n \subseteq R \cap (\mathscr{R}_0 \cap (\mathscr{F}_{R,I} + \mathscr{R}'))$$
$$= R \cap (\mathscr{F}_{R,I} + \mathscr{R}_0 \cap \mathscr{R}') = R \cap \mathscr{F}_{R,I} \subseteq R \cap Q_n.$$

It follows that $\ker \varphi = R \cap \mathscr{F}_{R,I}$ and, by part $2°$ of Proposition 4, condition c) of Theorem 2 implies that φ is an embedding. On the other hand, by Lemma 6 and part 2) of Corollary 1, the category $\mathscr{D}$ is representable. Finally, taking into consideration (17), we obtain the desired implication c) $\Rightarrow$ a) in the case when Φ is a field.

This implication can be proved directly for algebras over a regular commutative ring, but the proof becomes considerably more complicated. I. V. L'vov suggested a short method of reducing proofs of this type of statement from the case of algebras over a regular commutative ring to the case of algebras over a field. The essence of this method consists in the fact that quasi-identities of a ring R (which is an algebra over a regular commutative ring Φ) are transferred to the factor ring R/MR, where M is an ideal of the ring Φ. Modifying this method slightly we show how the implication c) $\Rightarrow$ a) is reduced to the case of algebras over a field.

Let the ground ring Φ be regular and let $\mathfrak{M}$ be the set of its maximal ideals; then the equality

$$\bigcap_{M \in \mathfrak{M}} (U + MR) = U \qquad (18)$$

holds for an arbitrary ideal U of the algebra R.

Indeed, assume that there exists an element x from the left-hand side of equality (18) that does not belong to U. Then the ideal $\{\alpha \in \Phi \mid \alpha x \in U\}$ of the ring Φ is proper, and hence it is contained in a maximal ideal M of Φ. But $x \in U + MR$ and, by the regularity of the ring Φ, we obtain $x = u + ev$, where $u \in U$, $v \in R$, and e is an idempotent from the ideal M. It follows that $(1-e)x = (1-e)u \in U$. Thus, $1-e, e \in M$, which contradicts the ideal M being proper. The opposite inclusion is obvious.

Now suppose that for every $M \in \mathfrak{M}$ there exists an embedding of algebras (over the field Φ/M) $\varphi^{(M)} \colon R/MR \to T_n(C_M)$ strictly compatible with the sequence $(I_1/MR, \ldots, I_n/MR)$ of ideals, where C_M is a commutative Φ/M-algebra. Observe that on every Φ/M-algebra A there is a naturally induced structure of a Φ-algebra $(\alpha \cdot x := (\alpha + M)x,\ \alpha \in \Phi,\ x \in A))$, where homomorphisms of Φ/M-algebras can be considered as homomorphisms of Φ-algebras. Therefore, we have a homomorphism

$$\psi \colon R \to T_n\left(\prod_{M \in \mathfrak{M}} C_M\right), \qquad \psi_{ij}(r) = (\varphi^{(M)}_{ij}(r + MR))_{M \in \mathfrak{M}},$$

of Φ-algebras, which, by equalities of the form (18), is an embedding strictly compatible with the sequence $(I_1, \ldots, I_n)$ of ideals.

Thus, it remains to show that if a Φ-algebra R with a sequence $(I_1, \ldots, I_n)$ of ideals satisfies condition c), then the Φ/M-algebra R/MR ($M \triangleleft \Phi$) with the sequence $\overline{I} = (I_1/MR, \ldots, I_n/MR)$ of ideals also satisfies condition c). Let g be an n-polynomial over Φ/M with variables from X, and let $\eta \colon X \to R/MR$ be an interpretation of variables such that $(\partial g/\partial w)\eta = 0$ for every nonempty monomial $w \in \mathfrak{X}(g, \eta, \overline{I})$. Let h denote an n-polynomial over Φ that is transformed into the polynomial g under the natural epimorphism $\Phi\langle X\rangle \to (\Phi/M)\langle X\rangle$. For each variable $x \in X_i$ ($i = 1, \ldots, n$) we choose an element $\xi(x) \in R$ such that $\eta(x) = \xi(x) + MR$ and if $\eta(x) \in I_i/MR$, then $\xi(x) \in I_i$. This determines an interpretation $\xi \colon X \to R$, where $\mathfrak{X}(h, \xi, I) \subseteq \mathfrak{X}(g, \eta, \overline{I})$. Therefore, $(\partial h/\partial w)\xi \in MR$ for every nonempty monomial $w \in \mathfrak{X}(h, \xi, I)$. Since the set $\mathfrak{X}(h, \xi, I)$ is finite, then by the regularity of the ring Φ, there exists an idempotent $e \in M$ such that $(\partial h/\partial w)\xi \in eR$ for every nonempty monomial $w \in \mathfrak{X}(h, \xi, I)$. It follows that the n-polynomial $f = (1-e)h$ over Φ and the interpretation ξ satisfy the antecedents of the implication (8). Therefore, it follows from condition c) for the algebra R with the sequence I of ideals that $f\xi = 0$, that is, $h\xi = eh\xi \in MR$. Thus, $g\eta = 0$. Theorem 2 is completely proved.

REMARK 5. Analogous arguments show that for algebras with identity in their signature Theorem 2 remains valid if, of course, the ideals $I_1, \ldots, I_n$ do not intersect the subalgebra of R generated by the identity. In particular, this implies the validity of Anan'in's conjecture for algebras with identity in their signature over a commutative regular ring.

§4. Representation of algebras with diagonal

It follows from Lemma 3 and Theorem 2 that every algebra over a field with diagonal d of order n that satisfies identity (11) is embeddable (strictly compatibly with the sequence $(\ker d_1, \ldots, \ker d_n)$ of ideals) in an algebra of upper triangular $(n \times n)$-matrices over a commutative algebra, although in general without preserving the diagonal. In this section we are going to consider cases when there exist such embeddings that preserve diagonal. It turns out that this problem has a positive answer for algebras with diagonal of order $n < 4$ that satisfy identity (11). On the other hand, as an example from the following section shows, for algebras with diagonal of order 4 this is false. Besides that, this problem is solved positively for algebras (with diagonal and satisfying identity (11)) with identity over a commutative regular ring in which the diagonal mappings act into the subalgebra generated by the identity. Here, in the last statement, one cannot drop the regularity of the ground ring.

For an F-algebra R let $UT_n(R)_F$ denote the subalgebra of the algebra of the upper triangular $(n \times n)$-matrices generated by all matrices that have elements of the ring F in the principal diagonal and elements of R above the main diagonal. The next lemma easily follows from Bergman's result on the representability of every unitriangular category we have already mentioned.

LEMMA 7. *For an arbitrary algebra R (over a commutative algebra F) there exists an embedding of F-algebras with diagonal $UT_n(R)_F \to UT_n(K)_F$, where K is a commutative F-algebra.*

THEOREM 4. *An algebra R with diagonal d of order $n < 4$ that satisfies identity (11) is embeddable with preservation of diagonal in the algebra of upper triangular $(n \times n)$-matrices over a commutative algebra.*

PROOF. The case $n = 1$ is trivial. Consider the case $n = 2$. Let F and M denote the image and the kernel of the homomorphism d_1, respectively, and let K be the F-module algebra of the F-module M, that is, $K = F \oplus M$ (here M is an ideal with zero multiplication in the F-algebra K). Taking into consideration the equality $d_1(R) = d_2(R)$, we easily obtain that the correspondence

$$r \mapsto \begin{bmatrix} d_1(r) & r - d_1(r) \\ 0 & d_2(r) \end{bmatrix}$$

determines the desired embedding of the algebra R into the algebra $T_2(K)$.

Case $n = 3$. Adjoining, if needed, an identity, one can ascertain that R

is an algebra with identity whose diagonal mappings preserve the identity. Let $d_{ii+1}\colon R \to R/(\ker d_i \ker d_{i+1})$ ($i = 1, 2$) be canonical epimorphisms of algebras and $K = (\operatorname{im} d_1)^{\#}$. We want to define a structure of an associative K-algebra on the K-module $\mathscr{R} = R \oplus \operatorname{im} d_{12} \oplus \operatorname{im} d_{23}$ in such a way that the mapping $\varphi\colon R \to UT_n(\mathscr{R})_K$ that transforms an element x into a matrix

$$\begin{bmatrix} d_1(x) & d_{12}(x) - d_1(x)e_1 - d_2(x)e_2 & x - d_{12}(x) - d_{23}(x) + d_2(x)e_2 \\ 0 & d_2(x) & d_{23}(x) - d_2(x)e_2 - d_3(x)e_3 \\ 0 & 0 & d_3(x) \end{bmatrix},$$

where $e_1 = 1 - d_{23}(1)$, $e_3 = 1 - d_{12}(1)$, $e_2 = 1 - e_1 - e_3$, is an embedding of algebras.

We point out a priori considerations that permit us to define the multiplication on the module $\mathscr{R}$. If x is an upper triangular (3×3)-matrix over a commutative ring, then $d_{ii+1}(x)$ would be considered as a matrix that is obtained from x by turning into zero all (k, l)-components such that $\{k, l\} \not\subseteq \{i, i+1\}$. Here, of course, the following equalities must hold: $0 = (x - d_{12}(x))(y - d_{23}(y)) = (x - d_1(x))(y - d_{23}(y)) = (x - d_{12}(x))(y - d_3(y))$, where d_i ($i = 1, 2, 3$) are natural diagonal mappings. Using these considerations, it becomes easy to write the multiplication table of the algebra $\mathscr{R}$ and to understand the meaning of the mapping φ.

Define the structure of a K-algebra on the module R compatible with the structure of a K-module and with multiplication in the algebras R, $\operatorname{im} d_{12}$, $\operatorname{im} d_{23}$ using these rules: $d_{23}(y)x = d_{23}(yx)$, $xd_{12}(y) = d_{12}(xy)$, $xd_{23}(y) = xy - d_1(x)(y - d_{23}(y))$, $d_{12}(x)y = xy - d_3(y)(x - d_{12}(x))$, $d_{23}(y)d_{12}(x) = d_2(yx)e_2$, $d_{12}(x)d_{23}(y) = xy - d_3(y)(x - d_{12}(x)) - d_1(x)(y - d_{23}(y))$.

Without difficulty we verify that this definition is correct and that the resulting multiplication is associative. Here we have, of course, to use identity (11) and also equality (12). Besides that, it is convenient to show first that e_1, e_2, e_3 are orthogonal idempotents.

A simple verification shows that φ is an embedding of algebras. It remains to use Lemma 7. Theorem 4 is proved.

As a corollary we easily obtain a description of T_n-algebras ($n = 2, 3$) over an arbitrary commutative ring with identity. First, we introduce some notation. Let $R_{(i)}$ ($i = 1, 2, 3$) denote the submodule of $R \otimes R \otimes R$ (R is an algebra) generated by all elements of the form $a_1 \otimes a_2 \otimes a_3$ ($a_1, a_2, a_3 \in R$, $a_i \in (R, R)$). Besides, if A and B are submodules in R, then $[A \otimes B]$ denotes the natural image of the module $A \otimes B$ in $R \otimes R$.

COROLLARY 5. 1°. *An algebra R is a T_2-algebra if and only if for every $a_i, b_i \in R$*

$$\sum_i a_i \otimes b_i \in [(R, R) \otimes R] \cap [R \otimes (R, R)] \Rightarrow \sum_i a_i b_i = 0.$$

2°. *An algebra R is a T_3-algebra if and only if for every $a_i, b_i, c_i \in R$*

the inclusions

$$\sum_i a_i \otimes b_i \otimes c_i \in (R_{(1)} + R_{(2)}) \cap (R_{(2)} + R_{(3)}) \cap (R_{(1)} + R_{(3)}),$$

$$\sum_i a_i \otimes b_i c_i, \quad \sum_i b_i \otimes a_i c_i, \quad \sum_i c_i \otimes a_i b_i \in [(R, R) \otimes R]$$

imply the equality

$$\sum_i a_i b_i c_i = 0.$$

PROOF. Indeed, by the previous theorem the algebra R is a T_n-algebra ($n = 2, 3$) if and only if it is embeddable in an algebra with diagonal of order n that satisfies identity (11). As we have already observed in §2, $R \cap \mathscr{F}_R = 0$ is necessary and sufficient for that. Now the statements of the corollary follow straightforwardly from the previous theorem and part 1° of Proposition 4.

THEOREM 5. *Let $\langle R, d \rangle_n$ be an algebra with identity over a commutative regular ring that satisfies identity* (11) *and where the diagonal mappings $d_1, \ldots, d_n$ act into the subalgebra generated by the identity. Then the algebra $\langle R, d \rangle_n$ can be embedded (preserving the diagonal) into the algebra of upper triangular $(n \times n)$-matrices over a commutative algebra.*

PROOF. 1°. Let the algebra R satisfy the conditions of the theorem, and let M be a maximal ideal of the ring Φ. Then on the Φ/M-algebra R/MR there is induced a structure of an algebra with diagonal $\overline{d}$ ($\overline{d}_i(r + MR) = d_i(r) + MR$) that satisfies identity (11). Now using arguments from the previous section, we can easily reduce the proof of the theorem to the case of algebras over a field.

2°. So let Φ be a field. Consider the filtration

$$R = W_0 \supseteq W_1 \supseteq \cdots \supseteq W_n = 0,$$

where $W_i = \ker \Delta_1 \cdots \ker \Delta_n$ ($\Delta_i(r) = r - d_i(r)$) and choose subspaces N_i in such a way that $W_{i-1} = N_i \oplus W_i$ ($i = 1, \ldots, n$). Since $R = \Phi \oplus W_1$ ($r = d_1(r) + \Delta_1(r)$), we may assume that $N_1 = \Phi$.

Define a unitriangular category $\mathscr{D}$ of order n (over Φ) by $\mathscr{D}(i, j) = \mathrm{Hom}_\Phi(N_i, N_j)$ ($i < j$); here the composition of morphisms is the superposition of linear mappings.

Arguing in the same way as in the proof of implication c) $\Rightarrow$ a) of Theorem 2, define linear mappings $r_{ij}: N_i \to N_j$ ($i \leq j$) by equalities $hr = hr_{ii} + hr_{ii+1} + \cdots + hr_{in}$ ($h \in N_i$).

Observe that it follows from the inclusions $N_i(r - d_i(r)) \subseteq W_i$ that the mapping r_{ii} coincides with the multiplication by the scalar $d_i(r) \in \Phi$.

It is easy to verify that the mapping $R \to M(\mathscr{D})$ ($r \mapsto (r_{ij})_{i \leq j}$) is an embedding of algebras with diagonal. It remains to use the representability of unitriangular categories. Theorem 5 is proved.

§5. Certain counterexamples

1. As it follows, for example, from Theorem 3, every nilpotent of index n algebra over a regular commutative ring is embeddable in the algebra of strictly upper triangular $(n \times n)$-matrices over a commutative algebra. The following example shows that for every commutative nonregular ring Φ with identity there exists a nilpotent of index $n > 3$ Φ-algebra that is not embeddable in any graded algebra A of the form $A = A_1 \oplus \cdots \oplus A_{2n-5}$.

Let $\alpha \in \Phi \backslash \alpha^2 \Phi$ be an arbitrary element without a regular inverse element. As R take the factor algebra of the algebra of polynomials over Φ of a single variable x without constant term modulo the ideal generated by the elements x^n and $\alpha x - x^{n-2}$. Let $\bar{x}$ denote the image of the variable x in the algebra R. Now, if $\varphi: R \to A_1 \oplus \cdots \oplus A_{2n-5}$ is a homomorphism into a graded algebra, then $\varphi(\alpha \bar{x}^2) = 0$. Indeed, let $\varphi(\bar{x}) = a_1 + \cdots + a_{2n-5}$ ($a_i \in A_i$); then $\alpha a_1 + \cdots + \alpha a_{2n-5} = \varphi(\alpha \bar{x}) = \varphi(\bar{x}^{n-2}) \in A_{n-2} \oplus \cdots \oplus A_{2n-5}$; whence, $\alpha a_1 = \cdots = \alpha a_{n-3} = 0$. Therefore, $\varphi(\alpha \bar{x}^2) = \alpha(a_1 + \cdots + a_{2n-5})(a_1 + \cdots + a_{2n-5}) = 0$.

It remains to show that $\alpha \bar{x}^2 \neq 0$. If this is not so, then

$$\alpha x^2 \equiv (\alpha x - x^{n-2})(\alpha_1 x + \cdots + \alpha_{n-2} x^{n-2}) \mod (x^n)$$

for certain $\alpha_1, \ldots, \alpha_{n-2} \in \Phi$. Equating the coefficients of x^2 and x^{n-1}, respectively (recall that $n > 3$), we obtain the equalities

$$\alpha = \alpha \alpha_1, \qquad 0 = \alpha \alpha_{n-2} - \alpha_1,$$

and it follows from them that $\alpha \in \alpha^2 \Phi$. This contradiction shows that R is the desired example.

2. It is interesting that in the previous example the algebra R is commutative. The following example shows that Anan'in's conjecture fails for algebras over a nonregular commutative ring. For simplicity, we produce an example of an algebra R that satisfies the identity $[x, y][z, t] = 0$ and is not a T_2-algebra.

Let, as above, α be an arbitrary element of a ring Φ that does not belong to $\alpha^2 \Phi$. Let F denote an algebra without identity that is freely generated by the variables x, y, and z, and take for R the factor algebra of the algebra F modulo the ideal U generated by the element $\alpha x - [y, z]$ and all monomials of degree four. It is easy to show that every homomorphism of the algebra R into a graded algebra $A_0 \oplus A_1$ (where A_0 is a commutative algebra) annihilates the element $\alpha x^2 + U$. Therefore, it suffices to prove that $\alpha x^2 \notin U$. Assume the opposite; then

$$\alpha x^2 \equiv \sum_{u,v} \beta_{u,v} u(\alpha x - [y, z]) v \mod F^4,$$

where u and v run over a finite set of words (which may be empty) of the letters x, y, z, $\beta_{u,v} \in \Phi$. Considering the coefficients x^2, xyz, and yzx, we obtain, respectively, the equalities $\alpha = \beta_{x,1}\alpha + \beta_{1,x}\alpha$, $0 = \alpha\beta_{1,yz} - \beta_{x,1}$, $0 = \alpha\beta_{yz,1} - \beta_{1,x}$ which imply the equality $\alpha = \alpha^2(\beta_{1,yz} + \beta_{yz,1})$ that contradicts the choice of the element α. Therefore, R is the desired example.

3. A counterexample to a question of A. Z. Anan'in. An algebra R is called centrally representable of order n if it is embeddable in a subalgebra of the algebra of upper triangular $(n \times n)$-matrices over a commutative algebra each matrix of which has the same elements in the main diagonal. In his dissertation, Anan'in suggested a certain conjecture a particular case of which can be formulated as follows: every algebra R (over a field) that is decomposable into a sum of its center and a certain nilpotent of index n ideal is centrally representable of order n. The following example shows that this is not true already for $n = 3$.

As R take an algebra (without identity) generated by variables x, t, y, z and with defining relations $[x, t] = [x, y] = [t, y] = [x, z] = [t, z] = zy = y^2 = z^2 = 0$, $t^3 = x^2 = yz$. It is clear that $R = Z + N$, where Z is the central subalgebra generated by the elements x and t, and N is the ideal generated by the elements y and z.

It is easy to verify that $N^3 = 0$ and that the elements x^3 and t^4 differ from zero. It remains to prove that every homomorphism φ of the algebra R into a graded algebra $A = A_0 \oplus A_1 \oplus A_2$ such that $A_0 \subseteq Z(A)$ annihilates one of the elements x^3 or t^4. To this end set $\varphi(r) = r_0 + r_1 + r_2$ ($r \in R$, $r_i \in A_i$). Since $(A, A) \subseteq A_2$, then $\varphi(x^2) = \varphi(t^3) = \varphi([y, z]) \subseteq A_2$, and hence, $x_0^2 = 0$, $2x_0 x_1 = 0$, $t_0^3 = 0$. If the characteristic of the field Φ differs from 2, then $x_0 x_1 = 0$ and $\varphi(x^3) = x_0^3 + 3x_0^2(x_1 + x_2) + 3x_0(x_1 + x_2)^2 + (x_1 + x_2)^3 = 0$. If the characteristic of the field Φ equals 2, then $\varphi(t^4) = t_0^4 + (t_1 + t_2)^4 = 0$. Q.E.D.

4. Returning again to nilpotent algebras, we produce an example that answers in the negative a question of I. V. L'vov: is a nilpotent of index n Φ-algebra R, where Φ is a commutative algebra over a field $\mathscr{k}$, embeddable as a Φ-algebra in the algebra of upper triangular $(n \times n)$-matrices over a commutative algebra? As we have already observed, such an algebra is embeddable as a $\mathscr{k}$-algebra in the algebra of strictly upper triangular $(n \times n)$-matrices over a commutative algebra.

The following example gives a negative answer to that question already for $n = 8$.

Let $\mathscr{k}$ be an arbitrary field, and let F be an algebra without identity freely generated by the variables t, x, y, z, u. Let the variable t have degree five and the remaining variables degree one. As R take the factor algebra of the algebra F modulo the ideal U generated by the elements $[y, z]^2$, $[t, y]$, $[t, z]$, $[t, x]$, $[t, u]$, $[y, z]u[y, z]$, $tx - [y, z]u^2[y, z]$, and all monomials of u, z, and u of degree not less than 7 and all monomials of degree not less than 8. The image of an element f under the canonical epimorphism $F \to R$ is denoted by $\overline{f}$. Let $\Phi = k[t]$. It is clear that R is a Φ-algebra $(tr := \overline{t}r, \ r \in R)$. By the composition lemma (see [5]) it is not complicated to verify that the element $t\overline{x}^2$ differs from zero. It remains to show that every homomorphism of Φ-algebras φ from R into the algebra of upper triangular matrices of degree 8 over a commutative algebra annihilates the element $t\overline{x}^2$.

The relations $[\overline{y}, \overline{z}]^2 = [\overline{y}, \overline{z}]\overline{u}[\overline{y}, \overline{z}] = 0$ ensure the equality to zero for all (i, j)-components $(j - i \leq 4)$ of the matrix $\varphi([\overline{y}, \overline{z}]\overline{u}^2[\overline{y}, \overline{z}]) = \varphi(t\overline{x}) = t\varphi(\overline{x})$. The last claim can be verified directly, but it follows also from the fact that in every T_4-algebra S for every $a, b \in (S, S)$, $v, w \in S$ the equalities $ab = 0$, $avb = 0$, $awb = 0$ imply the equality $awvb = 0$.

Let $\varphi(\overline{x}) = x_0 + x_1 + \cdots + x_7$, where x_i is a matrix with zeroes outside the ith diagonal over the main diagonal. Then the previous implies $tx_0 = tx_1 = tx_2 = tx_3 = 0$. Therefore, $\varphi(t\overline{x}^2) = t(x_0 + \cdots + x_7)(x_0 + \cdots + x_7) = 0$. Q.E.D.

5. In conclusion we produce an example of an algebra (over a field) with diagonal of order 4 that satisfies identity (11) and is not embeddable with preservation of diagonal into the algebra $T_4(K)$ for a commutative algebra K.

For simplicity, let Φ be a field of characteristic 2. Let R denote the commutative algebra generated by the variables t and x subject to the defining relations $x^4 = 0$, $t^4 x = x^2$. Define a diagonal of order 4 on R by setting $d_i(t) = t$, $d_i(x) = 0$ $(i = 1, 2, 3, 4)$. It is clear that the algebra R satisfies identity (11) for $n = 4$. It is easy to notice that the element $t^4 x^2$ differs from zero in the algebra R. Now let φ be a homomorphism of the algebra R into a graded algebra $A = A_0 \oplus A_1 \oplus A_2 \oplus A_3$ such that $A_0 \subseteq Z(A)$ and $\varphi(x) \in A_1 \oplus A_2 \oplus A_3$. Taking into account that $t^4 x = x^2$, we easily obtain the inclusion $\varphi(t^4)\varphi(x) \subseteq A_2 + A_3$. But $\varphi(t^4) \in A_0$ (because of the characteristic 2 of the ground field), and hence, $\varphi(t^4 x^2) = \varphi(t^4)\varphi(x)\varphi(x) = 0$.

Therefore, every homomorphism (preserving the diagonal) from the algebra R into the algebra of upper triangular matrices of order 4 over a commutative algebra annihilates the element $t^4 x^2 \neq 0$, and hence, is not an embedding.

Chapter 3
Special Representations of Nilpotent Graded Algebras

In this chapter we consider representations of graded algebras all of whose homogeneous components, except maybe from the first to the nth, equal zero.

A transitive relation T on the set of natural numbers is called a triangular configuration if

$$(i, j) \in T \Rightarrow i \leq j.$$

For an arbitrary algebra B and a triangular configuration T the ordinary "matrix" operations on $\bigoplus B_\pi$ ($B_\pi = B$, π runs over the set T) determine an algebra $T(B)$ of T-matrices over B. For example, for T-matrices $a = (a_\pi)_{\pi \in T}$, $b = (b_\pi)_{\pi \in T}$ over B and a pair (i, j) from T the equality

$$(ab)_{ij} = \sum_k a_{ik} b_{kj},$$

holds, where k runs over numbers such that (i, k), $(i, j) \in T$.

An algebra is called a T-algebra if it is embeddable in the algebra of T-matrices over a commutative algebra. In Proposition 5 we give a certain sufficient condition for T-representability. As a corollary we obtain that every $T_n^\varnothing$-algebra (in particular, every nilpotent of index n algebra over a field) for $n > 3$ is embeddable in the algebra of matrices of order $n-2$ over a commutative algebra, and this estimate cannot be improved.

§1. Embedding connected graded algebras

A graded algebra is called connected if it is generated by its first graded component.

A nondecreasing sequence $\pi = (\pi_i)_{i=0,\ldots,m}$ of natural numbers that does not contain three equal numbers is called a path of length m. Here T_π denotes a triangular configuration $\{(\pi_i, \pi_j) \mid 0 \leq i < j \leq m\}$. Also, we say that a graded algebra $R = R_1 \oplus \cdots \oplus R_n$ satisfies a π-quasi-identity (for a path π defined above) if, for every number t, k, and l such that $0 < t < m-1$, $\pi_t = \pi_{t+1}$, $1 \leq k \leq \min(t, n-2)$, $1 \leq l \leq \min(n-t-1, n-2)$, the following condition holds:

$$\sum_i x_i y_i = 0 \Rightarrow \sum_i x_i z y_i = 0 \qquad (x_i \in R_k, y_i \in R_l, z \in R).$$

The principal technical assertion of this chapter consists of the following proposition.

PROPOSITION 5. *Every connected graded nilpotent of index $n+1$ algebra that satisfies a π-quasi-identity for a path π of length n is a T_π-algebra.*

PROOF. Observe that every connected graded algebra, nilpotent of index $n+1$, is representable as a factor algebra of a free algebra (without identity) modulo a homogeneous ideal that contains all monomials of degree greater than n. On the other hand, every such factor algebra is a connected, graded, and nilpotent of index $n+1$ algebra.

Let U be a homogeneous ideal of a free algebra $F = \Phi\langle X \rangle^{\cdot}$ that contains all monomials of degree greater than n and such that the factor algebra F/U (with natural grading) satisfies a π-quasi-identity for a path $\pi = (\pi_i)_{i=0,\ldots,n}$.

We have to prove that the algebra F/U is a T-algebra for the triangular configuration $T = T_\pi$.

We assume that the pairs from T are lexicographically ordered, that is, $(i,j) < (k,l)$ if $i < k$ or $i = k$ and $j < l$. In the sequel we need the following notations: $\mathfrak{X}$ is a semigroup with identity freely generated by a set X, $\mathfrak{X}_m$ is the set of all monomials from $\mathfrak{X}$ of degree m, T_m is the set of all subsets of T of cardinality m. Besides, let $x_\theta = (x, \theta) \in X \times T$.

In the algebra of polynomials with identity of the set $X \times T$ of commutative variables set $x_\theta y_\theta = 0$ for all $x, y \in X$, $\theta \in T$. The resulting algebra is denoted by $\mathscr{L}$. For a monomial $a = x_{\theta_1}^{(1)} \cdots x_{\theta_m}^{(m)}$ ($x^{(i)} \in X$, $\theta_i \in T$, $\theta_1 < \cdots < \theta_m$) of the algebra $\mathscr{L}$ set $\tilde{a} = x^{(1)} \cdots x^{(m)}$. Conversely, for a monomial $u = x^{(1)} \cdots x^{(m)}$ from $\mathfrak{X}$ and a subset $\theta = \{\theta_1, \ldots, \theta_m\} \subseteq T$ ($x^{(1)}, \ldots, x^{(m)} \in X$, $\theta_1 < \cdots < \theta_m$) we set

$$u_\theta = x_{\theta_1}^{(1)} \cdots x_{\theta_m}^{(m)} \in \mathscr{L};$$

for $m = 0$ we assume that u_θ is the identity of the algebra $\mathscr{L}$.

To each subsequence $i_0, \ldots, i_k$ of π there corresponds a subset $\{(i_0, i_1), (i_1, i_2), \ldots, (i_{k-1}, i_k)\} \subseteq T$ which is called a path in T from i_0 to i_k of length k. Let τ be a path in T from i to j; then we set $\sigma(\tau) = (i, j)$. The set of all paths in T from i to j of length m is denoted by $T_{(i,j)}^m$.

For each pair θ from T define a linear mapping $F \to \mathscr{L}$ ($f \mapsto f_\theta$) by setting

$$u_\theta = \sum_{\tau \in T_\theta^m} u_\tau, \qquad (19)$$

for a monomial $u \in \mathfrak{X}_m$.

To avoid cluttering the notation with parentheses, we write f_{ij} instead of $f_{(i,j)}$ ($f \in F$, $(i,j) \in T$).

Obviously, for any elements u and v of the algebra F and a pair (i,j) from T we have the equality

$$(uv)_{ij} = \sum_k u_{ik} v_{kj},$$

where the summation is over all numbers k such that $(i,k), (k,j) \in T$.

Now, denoting by V the ideal of $\mathscr{L}$ generated by the set $\{f_\theta \mid f \in U, \ \theta \in T\}$, we easily convince ourselves that the mapping

$$F/U \to T(\mathscr{L}/V), \qquad f + U \mapsto (f_\theta + V)_{\theta \in T}$$

is defined correctly and is a homomorphism of algebras.

Now we prove that this homomorphism is injective. To this end it suffices to prove that

$$f_{\pi_0, \pi_m} \in V \Rightarrow f \in U$$

for every homogeneous of degree $m \leq n$ element f of the algebra F.

Let, for definiteness,

$$f = \sum_{u \in \mathfrak{X}_m} \lambda_u u \qquad (\lambda_u \in \Phi).$$

Then $f_\theta \in V$ ($\theta = (\pi_0, \pi_m)$) means that

$$f_\theta = \sum_{d=1}^{m} \sum_{\Delta \in P_d} g(\Delta)_\rho v_\tau, \qquad (20)$$

where the symbol Δ is introduced for brevity and denotes the sequence (ρ, τ, v), $P_d = T \times \mathscr{T}_{m-d} \times \mathfrak{X}_{m-d}$, and

$$g(\Delta) = \sum_{w \in \mathfrak{X}_d} C(\Delta, w) w \qquad (C(\Delta, w) \in \Phi)$$

are homogeneous polynomials from U of degree d.

It is clear that there exists a unique path in T from π_0 to π_m of length m, namely, the path $\eta = \{(\pi_0, \pi_1), (\pi_1, \pi_2), \ldots, (\pi_{m-1}, \pi_m)\}$. Therefore, writing the elements f_θ and $g(\Delta)_\rho$ by means of equality (19), we rewrite (20) in the form

$$\sum_u \lambda_u u_\eta = \sum_{d=1}^{m} \sum_\Delta \sum_w \sum_\delta C(\Delta, w) w_\delta v_\tau, \qquad (21)$$

where the elements u, Δ, w, δ run over the sets $\mathfrak{X}_m, P_d, \mathfrak{X}_d, T_\rho^d$, respectively.

To know how to collect similar terms in the right-hand side of this latter equality, we need the following definition.

Let δ, τ be disjoint subsets of T and u a monomial from $\mathfrak{X}$ whose length equals the cardinality of the set $\tau \cup \delta$. Then a (δ, τ)-decomposition of the monomial u is defined as a pair (w, v) of monomials from $\mathfrak{X}$ that is uniquely determined by the equality $u_{\delta \cup \tau} = w_\delta v_\tau$ in the algebra $\mathscr{L}$. Here we assume that $w = u(\delta, \tau)$, $v = u(\tau, \delta)$.

Also, put $\mathscr{P}(\eta) = \{\mu \subseteq \eta \mid \mu \text{ is a path in } T\}$.

Since the equality $u_\eta = w_\delta v_\tau$ ($w \in \mathfrak{X}_d$, $v \in \mathfrak{X}_{m-d}$, $\delta \in T^d_\rho$, $\tau \in \mathscr{T}_{m-d}$) holds if and only if

$$\delta \in \mathscr{P}(\eta), \quad \tau = \eta \backslash \delta, \quad w = u(\delta, \tau), \quad v = u(\tau, \delta),$$

then, equating the coefficients at u_η in the right- and left-hand sides of the equality (21), we obtain

$$\lambda_u = \sum_{\delta \in \mathscr{P}(\eta)} C(\Delta_{u,\delta}, u(\delta, \tau)),$$

where τ depends on δ, namely, $\tau = \eta \backslash \delta$; besides, $\Delta_{u,\delta} = (\sigma(\delta), \tau, u(\tau, \delta))$. Therefore,

$$f = \sum_u \sum_\delta C(\Delta_{u,\delta}, u(\delta, \eta \backslash \delta)) u, \qquad (22)$$

where the elements u and δ run over the sets $\mathfrak{X}_m$ and $\mathscr{P}(\eta)$, respectively.

For fixed $\delta \in \mathscr{P}(\eta)$, represent each monomial $u \in \mathfrak{X}_m$ in the form $\widetilde{w_\delta v}_\tau$, where $\tau = \eta \backslash \delta$ and the pair (w, v) of monomials is a (δ, τ)-decomposition of the monomial u, that is, $w = u(\delta, \tau)$, $v = u(\tau, \delta)$. Note that when the monomial u runs over the set $\mathfrak{X}_m$, the monomials w and v independently run over the sets $\mathfrak{X}_{|\delta|}$ and $\mathfrak{X}_{|\tau|}$, respectively.

This remark makes it possible to rewrite equality (22) in the form

$$f = \sum_\delta \sum_v \left(\sum_w C((\sigma(\delta), \eta \backslash \delta, v), w) \widetilde{w_\delta v}_{\eta \backslash \delta} \right), \qquad (23)$$

where elements δ, v, w run over the sets $\mathscr{P}(\eta)$, $\mathfrak{X}_{m-|\delta|}$, $\mathfrak{X}_{|\delta|}$, respectively.

Next we show that the right-hand side of equality (23) belongs to the ideal U. To this end we need the following

LEMMA 8. *Let* $\eta = \{(\pi_0, \pi_1), (\pi_1, \pi_2), \ldots, (\pi_{m-1}, \pi_m)\}$ *as above; then for every* $\delta \in \mathscr{P}(\eta)$, $\tau \subseteq \eta \backslash \delta$, $v \in \mathfrak{X}_{|\tau|}$ *the algebra* F *satisfies the condition*

$$\sum_w a_w w \in \mathscr{U} \Rightarrow \sum_w a_w \widetilde{w_\delta v}_\tau \in U,$$

where w *runs over the monomials from* $\mathfrak{X}_{|\delta|}$, $a_w \in \Phi$.

PROOF. The proof is by induction on the cardinality of the subset $\tau \subseteq \eta \backslash \delta$. Let $\tau = \{(i, j)\}$. There are two cases possible: $i = j$ and $i < j$. In the former case the statement of the lemma follows from the fact that the algebra F/U satisfies the π-quasi-identity. In the latter case the desired implication follows easily from the obvious equalities

$$\widetilde{w_\delta v}_\tau = \begin{cases} wv & \text{for } \sigma(\delta) < \tau, \\ vw & \text{for } \sigma(\delta) > \tau. \end{cases}$$

Now let the set τ contain more than one element. Then for suitable $\mu \subseteq \mathscr{P}(\eta)$, $\rho \in T$, $x \in X$, $y \in \mathfrak{X}_{|\mu|}$ we have the equalities $\tau = \mu \cup \{\rho\}$, $v_\tau = y_\mu x_\rho$. Using the equality

$$w_\delta \widetilde{y_\mu x_\rho} = \widetilde{(\widetilde{w_\delta y_\mu})_{\delta \cup \mu} x_\rho}$$

and the induction hypothesis ($|\mu| = |\tau| - 1$), we reduce the proof to the case when $|\tau| = 1$. Lemma 8 is proved.

Putting $\tau = \eta \backslash \delta$ in Lemma 8, we see that the inner sum in the right-hand side (and hence, the entire right-hand side) of equality (23) belongs to the ideal U. Proposition 5 is proved.

§2. Certain applications

We need the following

LEMMA 9. *Every graded algebra* $R = R_1 \oplus \cdots \oplus R_n$ *is embeddable with preservation of grading in a connected graded algebra* $C = C_1 \oplus \cdots \oplus C_n$.

PROOF. We may assume that $R = F/U$, where U is an ideal of the free algebra $F = \Phi\langle X\rangle\dot{}$ such that the image of each variable x from X under the natural epimorphism of algebras $F \to R$ differs from zero and belongs to a homogeneous component of the algebra R. Here if the image of the variable x belongs to the ith homogeneous component, we call the number i the weight of the variable x and denote it by $t(x)$. The weight of a monomial u of the algebra F is defined as the sum of weights of the variables from X that occur in u (counting multiplicities). Clearly, the ideal U is homogeneous with respect to the t-grading and contains all monomials of weight greater than n.

Let y be a new variable that does not belong to X. Extend the correspondence $x \mapsto xy^{t(x)-1}$ to a homomorphism of algebras

$$\varepsilon \colon F \to \mathscr{A} = \Phi\langle X \cup \{y\}\rangle\dot{}.$$

As C we take the factor algebra $\mathscr{A}/S$, where S is the ideal of the algebra $\mathscr{A}$ generated by the set $\varepsilon(U) + \mathscr{A}^{n+1}$. Since U is homogeneous with respect to weight, the ideal S is homogeneous with respect to degree. Therefore, the algebra C has the desired form.

Define a homomorphism of algebras $\lambda \colon R \to C$ transforming an element $f + U$ into the element $\varepsilon(f) + S$ ($f \in F$). This definition is correct since it follows from the inclusion $\varepsilon(\mathscr{U}) \subseteq S$. Obviously, the homomorphism λ preserves grading. Now we check that λ is an embedding.

Let $\lambda(f+U) = 0$ for some polynomial $f \in F$. This means that $\varepsilon(f) \in S$. Since the ideal U contains all monomials of weight greater than n and the ideal S all monomials of degree greater than n, we have

$$\varepsilon(f) \equiv \sum_{(v,w)} v\varepsilon(g_{v,w})w \mod \mathscr{A}^{n+1},$$

where $g_{v,w} \in U$ and the pair (v, w) of monomials (possibly empty) runs over a finite subset of $\mathscr{A}^{\#} \times \mathscr{A}^{\#}$. We split the sum on the right-hand side of the congruence into two sums. The first sum contains the summands for which $(v, w) \in \varepsilon(F)^{\#} \times \varepsilon(F)^{\#}$ and all others are in the second sum. Clearly, the first sum may be represented as $\varepsilon(g)$ for some $g \in V$, while the second sum is a linear combination of monomials from $\mathscr{A} \setminus \varepsilon(F)$. Therefore, $\varepsilon(f - g) \in \mathscr{A}^{n+1}$, that is, $f - g$ is a linear combination of monomials of weight greater than n. It follows that $f \in U$. Lemma 9 is proved.

A path $\pi = (\pi_i)_{i=0,\ldots,m}$ is called strict if $\pi_i < \cdots < \pi_{m-1}$. Also, we say that a path π runs in a triangular configuration T if $T_\pi \subseteq T$.

For a triangular configuration T let $t_1(T)$ (respectively, $t_2(T)$) denote the greatest length that a path (respectively, a strict path) running in T can have. If the lengths of paths in T are not bounded, we assume that $t_1(T) = t_2(T) = \infty$.

For brevity, let Γ_n denote the class of graded algebras all of whose components, except maybe from the first to the nth, equal zero.

Proposition 5 and Lemma 9 immediately imply

COROLLARY 6. *Every algebra from Γ_n is a T-algebra for every triangular configuration T such that $t_2(T) \geq n$.*

Put $T_n^0 = \{(i, j) \mid 1 \leq i < j \leq n\}$, $T_n^M = T_n^0 \cup \{(i, i) \mid i \in M\}$ ($M \subseteq \{1, \ldots, n\}$), $T_n = T_n^{\{1,\ldots,n\}}$.

Corollary 6 implies the following result due to Bergman [17] that every algebra from Γ_{n-1} is a T_n^0-algebra. To prove this it suffices to observe that $1, 2, \ldots, n-1, n$ is a strict path of length $n-1$ in T_n^0.

COROLLARY 7. *Every algebra in Γ_{n+1} is a $T_n^{\{1,n\}}$-algebra for $n \geq 2$ and a $T_{n+1}^{\{1\}}$-algebra for $n \geq 1$.*

Indeed, by Corollary 6, it suffices to observe that $1, 1, 2, 3, \ldots, n-1, n, n+1$ and $1, 1, 2, 3, \ldots, n-1, n, n$ are strict paths of length $n+1$ in the triangular configurations $T_{n+1}^{\{1\}}$ and $T_n^{\{1,n\}}$, respectively.

In fact, it can be easily proved that $t_2(T_n^{\{1,n\}}) = t_2(T_{n+1}^{\{1\}}) = n + 1$.

As we have already observed, a nilpotent of index $n \geq 2$ algebra over a field belongs to the class Γ_{n-1}. Therefore, we obtain from Corollary 7 that, in particular, every nilpotent of index $n > 3$ algebra over a field is representable by an algebra of triangular matrices of order $n - 2$ over a commutative algebra (this fact follows from Theorem 3 as well). On the other hand, the following example shows that there exist nilpotent of index $n > 3$ algebras not embeddable in the full algebra of matrices of order $\leq n-3$ over a commutative ring. As an algebra R take the algebra (without identity) with two generators x and y subject to the defining relations $xy^i x = 0$ ($i = 0, \ldots, n-4$), $w = 0$, where w is an arbitrary word in x and y of length n.

Clearly, $R^n = 0$ and the element $xy^{n-3}x$ differs from zero. However, every homomorphism from the algebra R into the algebra $M_m(K)$ ($m \leq n - 3$) of matrices of order m over a commutative algebra K annihilates the element $xy^{n-3}x$ (because the index m of the algebra $M_m(K)$ is algebraic over K). Therefore, R is the desired example.

We note another extreme case of Proposition 5.

COROLLARY 8. *Let* $R = R_1 \oplus \cdots \oplus R_n$ *be a connected graded algebra and let for all numbers* k, l ($1 \leq k, l \leq n - 1$) *the following condition hold*:

$$\sum_i x_i y_i = 0 \Rightarrow \sum_i x_i z y_i = 0 \quad (x_i \in R_k, \; y_i \in R_l, \; z \in R).$$

Then the algebra R *is a* T-*algebra for every triangular configuration* T *such that* $t_1(T) \geq n$.

The proof follows directly from Proposition 5 in which, as π, we take any path of length n in the triangular configuration T.

In particular, under the conditions of Corollary 8 for $n = 2m - 1$ the algebra R is a T_m-algebra, because $1, 1, 2, 2, \ldots, m, m$ is a path in T_m of length $2m - 1$.

This immediately implies

COROLLARY 9. *Let* P *be an ideal of a free algebra* $F = \Phi\langle Y \rangle'$ *such that* $F^{2m} \subseteq P \subseteq F^{2m-1}$. *Then the factor algebra* F/P *is a* T_m-*algebra*.

In particular, an algebra free in the variety of nilpotent of index $2n$ algebras is representable as an algebra of triangular matrices of order n over a commutative algebra. On the other hand, it is clear that an algebra free in the variety of nilpotent of index $2n$ algebras, which is not $2n$-generated, does not satisfy the standard identity of degree $< 2n$, and hence, is not embeddable in the full algebra of matrices of degree $< n$ over a commutative ring.

In conclusion we present a simple example showing that the quasi-identities in Corollary 8 cannot be omitted. Let F be the free algebra (without identity) of two generators x and y. The set

$$F^{2n} \cup \{xy^k x \mid k = 0, 1, \ldots, 2n - 4\} \quad (n > 2)$$

generates an ideal P. Then F/P is a connected graded algebra, nilpotent of index $2n$, and $t_1(T_n) = 2n - 1$. However, this algebra is not embeddable even in the full algebra of matrices of order $< 2n - 3$ over a commutative ring, because $xy^{2n-3}x \notin P$.

References

1. A. Z. Anan'in, *The imbeddings of algebras into algebras of triangular matrices*, Mat. Sb. **108** (1979), no. 2, 168–186; English transl. in Math. USSR-Sb. **36** (1979).
2. _____, *Representable varieties of algebras*, Algebra i Logika **28** (1989), no. 2, 127–143; English transl. in Algebra and Logic **28** (1989).
3. _____, *Locally finitely approximable locally representable varieties of algebras*, Algebra i Logika **16** (1977), no. 1, 3–23; English transl. in Algebra and Logic **16** (1977).
4. K. I. Beĭdar, *On matrix representation of endomorphism rings*, Abstracts of Reports, Fifth All-Union Symposium on Theory of Rings, Algebras and Modules (Novosibirsk, 1982), Inst. Math., Siberian Branch, 1982, p. 18. (Russian)
5. L. A. Bokut', *Associative Rings*, Part I, Novosibirsk. Gos. Univ., Novosibirsk, 1977. (Russian)
6. N. Bourbaki, *Algèbre*, Chaps. 1–3, Actualités Sci. Indust., nos. 934, 1236, 1044, Hermann, Paris, 1942, 1955, 1958.
7. S. I. Kublanovskiĭ, *Locally residually finite and locally representable varieties of associative rings and algebras*, Preprint No. 6143-82, deposited at VINITI, 1982. (Russian)
8. I. V. L'vov, *Representation of nilpotent algebras by matrices*, Sibirsk. Mat. Zh. **21** (1980), no. 5, 158–161; English transl. in Siberian Math. J. **21** (1980).
9. A. I. Mal'tsev, *On representations of infinite algebras*, Mat. Sb. **13** (1943), no. 2, 263–286. (Russian)
10. Yu. N. Mal'tsev, *A basis for the identities of the algebra of upper triangular matrices*, Algebra i Logika **10** (1971), no. 4, 393–400; English transl. in Algebra and Logic **10** (1971).
11. N. G. Nesterenko, *Representation of algebras by triangular matrices*, Algebra i Logika **24** (1985), no. 1, 65–86; English transl. in Algebra and Logic **24** (1985).
12. _____, *Representation by triangular matrices*, Abstracts of Reports, 18th All-Union Algebraic Conference (Kishinev, 1985), Part 2, p. 65. (Russian)
13. _____, *Quasi identities of T_n algebras*, Abstracts of Reports, 19th All-Union Algebraic Conference (L'vov, 1987), Part 2, p. 200. (Russian)
14. _____, *Triangular representations of nilpotent graded associative algebras*, Sibirsk. Mat. Zh. **29** (1988), no. 1, 122–128; English transl. in Siberian Math. J. **29** (1988).
15. Yu. P. Razmyslov, *Existence of a finite base for certain varieties of algebras*, Algebra i Logika **13** (1974), no. 6, 685–693; English transl. in Algebra and Logic **13** (1974).
16. C. Faith, *Algebra: Rings, Modules and Categories*. I, Grundlehren Math. Wiss., Band 190, Springer-Verlag, New York and Heidelberg, 1973.
17. G. M. Bergman, *Embedding rings in completed graded rings. I. Triangular embeddings*, J. Algebra **84** (1983), 14–24.
18. G. M. Bergman and S. M. Vovsi, *Embedding rings in completed graded rings. II. Algebras over a field*, J. Algebra **84** (1983), 25–41.
19. G. M. Bergman, D. J. Britten, and F. W. Lemire, *Embedding rings in completed graded rings. III. Algebras over general k*, J. Algebra **84** (1983), 42–61.
20. I. Kaplansky, *Problems in theory of rings*, Report of a Conference on Linear Algebras, June, 1956, Publ. 502, NAS-NRC, Washington, 1957, pp. 1–3.
21. _____, *"Problems in the theory of rings" revisited*, Amer. Math. Monthly **77** (1970), 445–454.
22. J. Lewin, *A matrix representation for associative algebras*. I, II, Trans. Amer. Math. Soc. **188** (1974), 293–317.
23. C. Procesi, *A formal inverse to the Cayley-Hamilton theorem*, J. Algebra **107** (1987), 63–74.

Translated by BORIS M. SCHEIN

PART III

Embedding Rings in Radical Rings and Rational Identities of Radical Algebras

A. I. Valitskas

Introduction

A leading role in the structure theory of associative rings is played by the Jacobson radical, which can be defined as the largest ideal in which every element is quasi-invertible, that is, possesses a two-sided inverse with respect to the adjoint multiplication $a \circ b = a+b-ab$. In the sequel, when discussing the radical of a ring R, we mean the Jacobson radical $J(R)$, unless stated otherwise.

In accordance with the general theory of radicals, two subclasses are singled out in the class of all associative rings: semisimple rings (those whose radical equals zero) and radical rings (which coincide with their radical). An arbitrary ring R is an extension of a radical ring $J(R)$ by a semisimple ring $R/J(R)$. Thus, the investigation of properties of a ring can be in many respects reduced to investigation of its radical and the semisimple quotient ring mentioned above.

By now methods of investigation of semisimple rings have been developed well enough; Jacobson's book "Structure of Rings" was trend-setting in this respect. However, up to now almost nothing is known about the structure of radical rings. Therefore, investigation of rings embeddable in radical ones may turn out to be useful for clarification of peculiarities of the structure of radical rings.

In 1947 Andrunakievich [1] proved a theorem containing sufficient conditions for embeddability of a ring in a radical ring; these are analogous to Ore's condition for embeddability of an integral domain in a division ring. He also raised the problem of finding necessary and sufficient conditions of embeddability of an arbitrary ring into a radical one [15, Problem 1.7].

From the point of view of universal algebra this embeddability problem belongs to the problems of representation of one axiomatizable class of algebraic systems in another class. Also, the class of rings embeddable in radical rings forms a quasivariety, and hence can be defined by a system of quasi-identities (universal Horn sentences). Recall that an abstract class of algebraic systems is a quasivariety if and only if it contains the singleton system and is closed both under subsystems and filtered products. As other examples of quasivarieties, we mention classes of semigroups embeddable in groups and subrings of direct products of division rings.

A system of quasi-identities that determines the class of semigroups embeddable in groups was found by Mal′tsev [20] in 1939. He produced examples of integral domains not embeddable in rings and raised the problem of describing those rings for which such an embedding is possible [15, Problem 1.103]. An answer to this problem was given in 1971 by Cohn [16].

Not only do these works of Mal′tsev solve a concrete problem of representation of semigroups in groups, but they chart a wide research program in this field. We recall just a few of its magisterial directions, restricting ourselves to the consideration of results directly connected with the subject of this dissertation.

§1. Building universal constructions

One of the methods for obtaining representability conditions of one axiomatizable class of algebraic systems in another is connected with designing universal constructions. In general, and with no claim of absolute rigor, the method can be described as follows.

Suppose that R is an algebraic system of signature τ, $\mathfrak{R}$ a class of algebraic systems of an extended signature $\theta \supseteq \tau$, and $\mathscr{H}$ a fixed class of homomorphisms from R into algebraic systems from $\mathfrak{R}$. Let there exist a homomorphism $\sigma: R \to U(R)$ from $\mathscr{H}$ that possesses the following universal property: for every homomorphism $f: R \to S$, $f \in \mathscr{H}$ there exists a uniquely determined homomorphism $\bar{f}: U(R) \to S$ that completes a diagram of the form

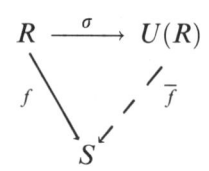

$$\begin{array}{ccc} R & \xrightarrow{\sigma} & U(R) \\ & {\scriptstyle f}\searrow & \downarrow {\scriptstyle \bar{f}} \\ & & S \end{array}$$

to a commutative one. Clearly, $\mathscr{H}$ contains an injective homomorphism if and only if the universal homomorphism $\sigma: R \to U(R)$ possesses this property. Moreover, if the construction of this universal homomorphism is sufficiently suitable for computations (for example, if the kernel congruence connected with σ can be described in terms of R), then the condition of embeddability of R into an algebraic system from $\mathfrak{R}$ by a homomorphism from $\mathscr{H}$ can also be written in terms of R.

Universal constructions for these concrete problems are connected with inversion of elements. Let R be a semigroup (a ring), and let Σ be an arbitrary subset of R. A semigroup (ring) homomorphism $f: R \to S$ is called Σ-inverting if the image of any element of Σ under f is invertible in S. An analogous definition can be given in the case when Σ is an arbitrary set of matrices over a ring R. If there exists an injective Σ-inverting homomorphism, the set Σ of elements (matrices) of a semigroup (over a ring) R is called potentially invertible. In the same way one defines Σ-quasi-invertible homomorphisms of a ring and potentially quasi-invertible sets of matrices.

Taking an arbitrary semigroup for R and the classes of all groups and all R-inverting semigroup homomorphisms for $\mathfrak{R}$ and $\mathscr{H}$, respectively, in the above scheme, we arrive at the notion of a universal R-group, whose construction, suggested by Mal'tsev, led to a description of the class of semigroups embeddable in groups.

The problem of finding conditions for embedding rings in division rings required additional efforts because even if one repeats the operation of inverting all nonzero elements of a ring several times, one may not arrive at a division ring. Cohn's discovery consisted in using inversion of matrices to obtain division ring constructions. Namely, he showed in [16] that homomorphisms of rings into division rings are determined (up to division ring embeddings) by sets of matrices that become invertible under these homomorphisms. Moreover, such sets of matrices, called localizing by Cohn, can be described in terms of the original ring. If a ring R and a localizing set Σ of matrices are fixed, the class of all division rings chosen for $\mathfrak{R}$ and $\mathscr{H}$ is the set of all ring homomorphisms which turn matrices from Σ into invertible ones, then the general scheme leads us to the concept of a universal Σ-inverting R-division ring, whose construction Cohn [16] was able to find.

The concept of a universal Σ-inverting ring $R\Sigma^{-1}$, also introduced by Cohn, is connected with inverting elements in arbitrary rings (and not necessarily in division rings). A corresponding definition is obtained if Σ is an arbitrary set of matrices over R, $\mathfrak{R}$ is the class of all rings, and $\mathscr{H}$ is the set of all Σ-inverting homomorphisms. The importance of a construction of the ring $R\Sigma^{-1}$ useful for calculations lies in the possibility of transferring localization techniques to noncommutative rings. Such a construction and its numerous applications were found by Gerasimov [12] and [13] and later by Malcolmson [40] and [41].

A construction of a universal Σ-inverting ring $R\Sigma^{-1}$ can be used to solve the above-mentioned problem of Andrunakievich on necessary and sufficient conditions for embeddability of rings in radical rings. Namely, Cohn [32] and later the author [7] proved that the ring $J(R^{\#}\Sigma^{-1})$ (where $R^{\#}$ is a ring obtained from R by adjoining an external identity and Σ is a set of matrices that become invertible under the natural projection $R^{\#} \to R^{\#}/R \simeq \mathbb{Z}$) is a universal radical R-ring. Thus, a ring R is embeddable in a radical one if and only if the universal homomorphism $\sigma\colon R^{\#} \to R^{\#}\Sigma^{-1}$ is an embedding.

§2. Investigation of complexity of quasivarieties of algebraic systems

As we have observed above, the class of semigroups embeddable in groups and also the classes of rings embeddable in direct products of division rings and in radical rings, respectively, are quasivarieties. There arises a natural problem on complexity of their determining systems of quasi-identities. In more concrete terms, we consider the following problem: can a given quasivariety be determined by a finite system of quasi-identities?

This problem was raised and solved in the negative for the quasivariety of semigroups embeddable in groups in Mal'tsev's article [20]. An analogous result for the quasivariety of rings embeddable in direct products of division rings was obtained by Gerasimov [11] and, independently, by Cohn [34].

In both cases the method of proof consists of designing (for every natural n) classes of semigroups (rings) such that none of them belongs to the quasivariety considered, but the intersection of these classes (over all n) does. Thus, the quasivariety considered is compared with certain similar classes of algebraic systems.

Many known questions are formulated in the language of such a comparison. Namely, let $\mathscr{K}_0$ be the class of rings whose multiplicative semigroup satisfies the two-sided cancellativity condition:

$$\forall x \forall y \forall z \, ((xy = xz) \vee (yx = zx) \Rightarrow (y = z));$$

let $\mathscr{K}_1$ be the class of rings whose multiplicative semigroup is embeddable in a group; $\mathscr{K}_2$ the class of invertible rings (those for which the set $R \setminus \{0\}$ is potentially invertible); $\mathscr{K}_3$ the class of rings embeddable in division rings. Clearly, the following inclusions hold: $\mathscr{K}_0 \supseteq \mathscr{K}_1 \supseteq \mathscr{K}_2 \supseteq \mathscr{K}_3$. The question of whether $\mathscr{K}_0$ and $\mathscr{K}_3$ coincide is a reformulation of van der Waerden's problem solved by Mal'tsev in his article [20] which we cited. Moreover, an example constructed by Mal'tsev shows that $\mathscr{K}_0 \neq \mathscr{K}_1$.

The question of whether $\mathscr{K}_1$ and $\mathscr{K}_3$ coincide is known as Mal'tsev's problem (see [21, p. 6]). This problem was solved independently by Bowtell [27], Klein [37], and Bokut' [4], [5], and [6]. The example of Bokut' is remarkable for two reasons. First of all, it shows that $\mathscr{K}_1 \neq \mathscr{K}_2$ (the examples of Bowtell and Klein are contained in $\mathscr{K}_2$), secondly, an algebra over the two-element field constructed by Bokut' is the only known semigroup algebra with this property. The problem on the existence of examples of algebras over a field of arbitrary characteristic that show that $\mathscr{K}_1 \neq \mathscr{K}_2$ is formulated in [15, Problem 2.22].

Finally, the inequality $\mathscr{K}_2 \neq \mathscr{K}_3$ is a corollary to results of Gerasimov [13] who proved that every 2-FI-ring is invertible, and to the existence of 2-FI-rings not embeddable in division rings (examples of Bowtell and Klein). Recall that a ring is called an n-FI-ring if each of its n-generated right ideals is a free right module of a uniquely determined rank.

Analogous problems appear in studies of embedding of rings in radical rings (all needed properties of the multiplicative semigroup of a ring should be replaced by corresponding properties of its adjoined semigroup). An example that proves the existence of a ring whose adjoined semigroup is cancellative by not embeddable in a group is analogous to Mal'tsev's example and is given by Andrunakievich [1].

§3. Investigation of when concrete classes of algebraic systems belong to a given quasivariety

Although the general form of the quasi-identities determining quasivarieties we investigate may be fairly simple, verification of the validity of these quasi-identities for concrete algebraic systems is, as a rule, very difficult. This is why proving that sufficiently wide classes of algebraic systems belong to a given quasivariety is so important. This type of problem is closely connected to problems of the preceding direction (then a quasivariety was compared with external classes and now with the internal ones). For example, when proving the theorem on the absence of a finite base of quasi-identities for the quasivariety of semigroups embeddable in groups, Mal'tsev [20] studied semigroups determined by defining relations of the form $x_i y_j = x_k y_l$ and asked when such semigroups are embeddable in groups. The works of L. A. Bokut' on groups with a standard base are a development of these ideas of Mal'tsev. Other important classes of semigroups embeddable in groups appeared in known theorems of R. Doss, S. I. Adyan, and A. Klein. ([1])

In exactly the same way, if one considers the class of rings embeddable in division rings, the groundbreaking theorems belong to O. Ore, A. I. Mal'tsev, and B. Neumann (on embedding group algebras of ordered groups in division rings) and P. M. Cohn (on valuation rings, which implies, in particular, that the universal enveloping algebra of an arbitrary Lie algebra is embeddable in a division ring). Also, an important role is played by Cohn's theorem on embedding semi-FI-rings (that is, rings that are n-FI-rings for all n) in division rings. Applications of this theorem form a significant part of Cohn's book [35].

From this point of view the paper of Andrunakievich [1], in which it was shown that Ore's conditions and two-sided cancellativity in the adjoined semigroup of a ring suffice for its embeddability in a radical ring, initiated constructions of similar classes of rings embeddable in radical ones.

In Cohn's article [31] sufficient conditions for embeddability of a ring in a simple radical ring were found.

The problem of finding necessary and sufficient conditions for embeddability of the universal envelope of an arbitrary Lie algebra in a radical ring was raised by L. A. Bokut'.

Finally, we mention yet another important direction in the modern ring theory (it does not have direct bearing to Mal'tsev's papers [20]): identities in an extended signature (type). A connection of this direction with the class of radical rings is explained by the fact that the latter class is a variety in the signature that consists of the ring signature together with the unary operation of quasi-inversion.

([1])[Translator's footnote]. See also papers by D. Tamari.

As with every variety of algebraic systems, this variety of radical rings possesses free systems, and hence, it makes sense to consider identities that hold on rings from this variety. We call such identities in extended signature rational by analogy with rational identities introduced by Amitsur [24] for the class of division rings. Speaking informally, a rational identity of a division ring D is any rational expression (an expression obtained from variables by means of the operations of addition, multiplication, subtraction, and formal inversion) which takes the zero value under every substitution of elements of D for variables such that all inversion operations are defined; a rigorous definition of this concept is given in Chapter 4 of this dissertation. A rational identity of D is called nontrivial if it does not hold in at least one division ring.

The main result proved by Amitsur [24] can be stated as follows: every division ring with an infinite center and satisfying a nontrivial rational identity is finite-dimensional over the center. In particular, such a division ring satisfies a polynomial identity (that is, it is a PI-ring). Later the proof of this fact was simplified by Bergman [26] and Cohn [30]. Gerasimov [15, Problem 2.31] asked whether a radical ring satisfying a nontrivial rational identity is a PI-ring.

The goal of this dissertation is obtaining answers to the problems mentioned above.

All rings considered are assumed to be associative, unless indicated otherwise. The existence of an identity in a ring is mentioned specifically in the instances in which this is not clear from the context. Moreover, if a ring has an identity, we assume that this identity is preserved under homomorphisms and, if it differs from zero, it acts identically in modules, although it does not have to be inherited by subrings. Also the natural extension of a ring homomorphism onto sets of matrices is denoted by the same letter as the homomorphism itself. The list of the most frequently used notations is given on page 137.

The dissertation consists of four chapters. The last three chapters are independent from each other, however, all of them use to some extent the notation, concepts, and results of Chapter 1. Each of the chapters has its own numeration of formulas, lemmas, propositions, and theorems. Here the number of every statement consists of two digits, the first of them is the number of the chapter, while the second one is the ordinal number of this statement in the chapter.

Chapter 1 is devoted to a proof of the following result:

THEOREM 1.7. *The quasivariety of rings embeddable in radical rings cannot be defined by a finite system of quasi-identities.*

In all arguments the fundamental role is played by a construction of matrix localization due to Gerasimov [13]. For the reader's sake this important construction is produced without proof in the first section of the chapter.

In the second section we consider problems whose study was initiated in Cohn's article [33]. Namely, methods developed by Gerasimov are applied to investigating a localization, with respect to a set Σ of matrices, of a ring S with identity that satisfies the following property: there exists an ideal R of S such that the natural projection $S \to S/R$ is Σ-inverting. In particular, we describe the kernel of the homomorphism $R \to S\Sigma^{-1}$ in terms of quasi-identities. In the case when $S = R^{\#}$ and Σ is the set of all matrices that become invertible under our projection, this system of quasi-identities obtained determines a quasivariety of rings embeddable in radical rings (Corollary 2 to Theorem 1.4).

In the third section of the chapter we consider certain sufficient conditions for potential invertibility of a set of matrices. The classes of rings introduced in that section generalize the n-FI-rings of Cohn. Another particular case of the rings considered, the class of A_n-rings, plays the same role in the investigations of ring embeddings in radical rings as the class of n-FI-rings for embeddings in division rings. Namely, we prove that a set of square matrices of order n over an A_{2n}-ring is potentially quasi-invertible (Proposition 1.2). In particular, a ring which is an A_n-ring for every natural n is embeddable in a radical ring.

Examples of A_n-rings produced in the fourth section show the existence of potentially quasi-invertible rings that are not embeddable in radical rings, and complete our proof of the main result of Chapter 1.

Chapter 2 is devoted to construction of examples of not invertible (not quasi-invertible) algebras over an arbitrary field whose multiplicative (adjoined) semigroup is embeddable in a group.

Namely, we consider an algebra R with identity over an arbitrary field, presented by defining relations of the following form:

$$\left(\begin{array}{c|c} a' & \tilde{a} \\ \hline a^0 & 'a \end{array}\right) \cdot \left(\begin{array}{c|c} b' & \tilde{b} \\ \hline b^0 & 'b \end{array}\right) = \left(\begin{array}{c|c} 0 & * \\ \hline 0 & 0 \end{array}\right),$$

where in the left-hand side general matrices are written with a selected partition into blocks whose product can be found according to the rules of operations with block matrices; here a^0 and b^0 are square upper-triangular matrices and the order of each of them is no less than two, and the asterisk denotes elements whose values are not restricted. It turns out that R is the desired example of a noninvertible algebra whose multiplicative semigroup is embeddable in a group.

In the first section of the chapter the structure of the multiplicative semigroup of a completion of a ring R is studied. Observe that a description of that semigroup can be given in terms of generators and defining relations, analogously to the description of multiplicative semigroups of $\overline{SN}$-rings introduced by Bokut' [3] in his solution of Mal'tsev's problem. Since we do not need such a description in full, we give only a proof of a theorem equivalent to Theorem 2 of [3.II] (Theorem 2.1 of the dissertation). The proof itself is a simplified version of arguments of Bokut' that use concepts from Chapter 1 of the dissertation and is given for completeness' sake.

By means of Theorem 2.1 the semigroup studied is split into a free product of two subsemigroups with an amalgamated subgroup (Theorem 2.2). The second section of this chapter is dedicated to a proof of embeddability in a group of each of the free factors obtained. Here we use a theorem of Doss (on embedding a rigid cancellative semigroup in a group) and ideas from a paper [17] by Koshevoĭ. As a corollary to previous considerations, we construct here an example of an algebra that is not quasi-invertible and whose adjoined semigroup is embeddable in a group.

In Chapter 3 of the dissertation a partial answer is given to a question of Bokut' on representations of Lie algebras in radical rings. Namely, we prove

THEOREM 3.1. *The following conditions are equivalent for a finite-dimensional Lie algebra*:

(a) *The universal envelope $U(L)$ without identity of the algebra L is embeddable in a radical ring.*

(b) *There exists an embedding of Lie rings $L \to R^{(-)}$ into an adjoined Lie ring of a radical ring R.*

(c) *The Lie algebra L is solvable.*

The principal part of the proof of this theorem is a verification of the embeddability of the universal envelope of a solvable (not necessarily finite-dimensional) Lie algebra in a radical ring. Results of Chapter 1 of the dissertation reduce this problem to that of potential invertibility of the set of matrices of the form $1 + A$, where A is an arbitrary square matrix over $U(L)$. The solvability condition for L makes it possible to conclude that $U(L)$ is an Ore domain [22]. Therefore, it possesses a division ring of quotients over which every matrix which is not a zero divisor is invertible. Verification of the fact that matrices of the form $1 + A$ are not zero divisors in the matrix ring over $U(L)^{\#}$ concludes the proof of Theorem 3.1.

By means of Theorem 3.1 one establishes the undecidability of the word problem in the class of radical rings, proved for the class of division rings by Macintyre [38] and [39].

Finally, radical algebras satisfying a nontrivial rational identity are studied in Chapter 4. It turns out that the following theorem holds:

THEOREM 4.1. *Let R be a radical algebra over an infinite field F that satisfies a nontrivial rational identity. Then R is a* PI-*algebra.*

The idea of the proof of this fact is close to Amitsur's article [24] and is given in the second section of the chapter.

The key to the proof of Theorem 4.1 is given by the following Theorem 4.2 that is proved in the third section.

THEOREM 4.2. *Let R be a prime radical algebra over an arbitrary field that satisfies a nontrivial rational identity. Then R possesses a classical ring* *of quotients that is isomorphic to the full matrix ring of certain order* m *over a division ring* D *(that is, it is an order in* D_m*)*.

The argument used to obtain this result is close to the known proof of Kaplansky's theorem on primitive PI-algebras that uses Jacobson's density theorem. Thus, assuming that a prime radical algebra that satisfies a nontrivial rational identity is not a Goldie ring, we are able to prove that it possesses a uniform right ideal. This makes it possible to conclude, applying Amitsur's general density theorem, that such an algebra is a dense ring of linear transformations of a torsion-free module over an Ore domain. The existence of a nontrivial rational identity in such an algebra leads easily to the desired contradiction.

The first section of the chapter is of technical character. Certain special kinds of elements of a free radical algebra are introduced there. Here all computations are based on the results of Chapter 1 of the dissertation.

The main results of the dissertation have been published in the author's works [7]–[10] and reported at the seminars "Algebra and logic", "Theory of rings", and "Associative rings and Lie rings" of the Institute of Mathematics of the Siberian Branch of the USSR Academy of Sciences and the Novosibirsk State University, and also at the Fifth All-Union Symposium on the Theory of rings, algebras, and modules (Novosibirsk, 1982).

Using this opportunity, I express sincere gratitude to my scientific advisor Professor L. A. Bokut'; scientific communication with him has largely defined the themes of this dissertation, and the all-encompassing help and support rendered by him had a great influence on its writing.

I thank also V. N. Gerasimov, whose works produced an immense impression on me, for his kindness and valuable advice.

Index of Notation

$\mathbb{N}$ is the ring of natural numbers.
$\mathbb{Z}$ is the ring of integers.
$F\langle \mathfrak{X} \rangle$ is the associative F-algebra with the set $\mathfrak{X}$ of free generators.
$F\langle\langle \mathfrak{X} \rangle\rangle$ is an algebra of formal power series of the elements of the set $\mathfrak{X}$.
$F_L\langle \mathfrak{X} \rangle$ is a free Lie F-algebra with the set $\mathfrak{X}$ of free generators.
$\circ$ is the adjoint multiplication operation: $a \circ b = a + b + ab$.
$J(R)$ is the Jacobson radical of a ring R.
R^k is the left R-module of rows of length k over a ring R.
$^k R$ is the right R-module of columns of length k over a ring R.
$^m R^n$ is the set of $(m \times n)$-matrices over a ring R.
R_n is the ring of square matrices of order n over a ring R.
$R_\infty = \bigcup_{n=1}^\infty R_n$.
$1 + R_\infty = \{1 + A \mid A \in R_\infty\}$.
$\mathrm{GL}(R)$ is the set of all (rectangular) matrices invertible over R (a rectangular $(m \times n)$-matrix a is called invertible over a ring R if there exists a rectangular matrix a^{-1} over R of size $n \times m$ such that $a^{-1}a = 1_n$ and $aa^{-1} = 1_m$, where the indices show the orders of the corresponding identity matrices.
$\mathrm{GL}_n(R) = \mathrm{GL}(R) \cap R_n$.
$R^\#$ is the ring obtained from R by adjoining an external identity.
R^* is the semigroup of nonzero elements of an integral domain R (the multiplicative semigroup of a ring R).
$R^{(-)}$ is the adjoined Lie ring of an associative ring R.
$R[t]$ is the ring of polynomials of a (commuting) variable t with coefficients from R.
$R[[t]]$ is the ring of formal power series of a (commuting) variable t with coefficients from R.
$R\Sigma^{-1}$ is the universal Σ-inverting ring.
$U_J(R)$ is the universal radical R-ring.

$$U_J\langle \mathfrak{X} \rangle = U_J(F\langle \mathfrak{X} \rangle).$$

$\mathrm{mult}(R, \Sigma)$ is the multiplicative closure of a set Σ of matrices over a ring R.

$U(L)$ is the universal envelope of a Lie algebra L.

$[L, L]$ is the commutant of a Lie algebra L.

$I(R)$ is the set of rational identities of a radical algebra R.

$\text{ann}_r(M) = \{x \in R \mid Mx = 0\}$ is the right annihilator of a set M.

$I \leq^{\text{ess}} M$ means that I is an essential submodule of a module M, that is, it has a nonzero intersection with every nonzero submodule in M.

$Z_r(M) = \{x \in M \mid \text{ann}_r(x) \leq^{\text{ess}} R\}$ is the right singular submodule of a module M.

$M \oplus N$ is the direct sum of R-submodules.

$U \dotplus V$ is the direct sum of vector spaces.

$d(M)$ is the Goldie dimension of an R-module M (the supremum of the numbers of nonzero summands in direct sums of submodules from M).

Chapter 1
Absence of a Finite Basis of Quasi-identities for the Quasivariety of Rings Embeddable in Radical Rings

This chapter is devoted to a proof of the result formulated in its title (Theorem 1.7). In all arguments the construction of matrix localization suggested by Gerasimov [13] plays a fundamental role. For the readers' sake this important construction is presented (without proofs) in the first section.

In the second section we consider topics whose study was initiated by Cohn [33]. Namely, methods developed by Gerasimov are applied to the investigation of the localization of a ring S with identity with respect to a set Σ of matrices satisfying the following property: there exists an ideal R of the ring S such that the natural projection $S \to S/R$ is Σ-inverting. In particular, we describe the kernel of the homomorphism $R \to S\Sigma^{-1}$ in terms of quasi-identities. In the case when $S = R^{\#}$ and Σ is the set of all matrices that are mapped into invertible ones by this projection, the system of quasi-identities we obtain determines a quasivariety of rings embeddable in radical rings (Corollary 2 to Theorem 1.4).

In the third section of the chapter we consider certain sufficient conditions for potential invertibility of a set of matrices. The classes of rings introduced in this section are generalizations of Cohn's n-FI-rings. Another particular case of the concepts considered, the class of A_n-rings, fulfills the same function in the studies of embedding rings in radical rings as the class of n-FI-rings does for embeddings in division rings. Namely, we prove that the set of square matrices of order n over an A_{2n}-ring is potentially invertible (Proposition 1.2). In particular, a ring that is an A_n-ring for any natural n is embeddable in a radical one.

Examples of A_n-rings given in the fourth section show the existence of potentially quasi-invertible rings that are not embeddable in radical ones, thus completing the proof of the principal result.

§1. A construction of matrix localization suggested by V. N. Gerasimov

Let R be an arbitrary ring with identity, and let Σ be a set of rectangular matrices over R.

Recall that a ring $R\Sigma^{-1}$ together with a homomorphism $\sigma\colon R \to R\Sigma^{-1}$ is called *universal Σ-inverting* if, for every ring homomorphism $f\colon R \to S$ that maps matrices from the set Σ into matrices invertible over S, there exists a uniquely determined homomorphism $\overline{f}\colon R\Sigma^{-1} \to S$ which completes the diagram of the following form up to a commutative one:

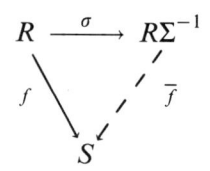

$$\begin{array}{ccc} R & \xrightarrow{\sigma} & R\Sigma^{-1} \\ & \searrow f & \swarrow \overline{f} \\ & S & \end{array}$$

Obviously, for every ring R and set Σ of matrices over R, the universal Σ-inverting ring $R\Sigma^{-1}$ exists, and it is uniquely determined up to isomorphism.

In the case when the homomorphism $\sigma\colon R \to \Sigma^{-1}$ is injective, the set Σ is called *potentially invertible*.

A set Σ is called *multiplicative* if it satisfies simultaneously the following two conditions:

(1) the empty (0×0)-matrix belongs to Σ;
(2) if $a, b \in \Sigma$, then $\begin{pmatrix} a & c \\ 0 & b \end{pmatrix} \in \Sigma$ for any matrix c of a suitable size over R.

Obviously, for every set Σ of rectangular matrices over a ring R the set

$$\mathrm{mult}(R, \Sigma) = \left\{ \begin{pmatrix} a_1 & & * \\ & \ddots & \\ 0 & & a_n \end{pmatrix} \mid a_i \in \Sigma \ (0 \leq i \leq n) \right\}$$

of matrices, where $*$ denotes arbitrary matrices of suitable sizes over R, is the least multiplicative set that contains Σ. It is called the *multiplicative closure* of the set Σ over the ring R.

Clearly, the universal inverting ring for the sets Σ and $\mathrm{mult}(R, \Sigma)$ of matrices are isomorphic. Therefore, in the sequel, unless otherwise stipulated, we assume that the set Σ is multiplicative.

Now we give (without proof) an explicit construction of the ring $R\Sigma^{-1}$ given by Gerasimov in [13].

Let $\mathscr{M}(R, \Sigma)$ denote the set of rectangular matrices a over R with a fixed block partition

$$a = \left(\begin{array}{c|c} a' & \tilde{a} \\ \hline a^0 & 'a \end{array} \right), \tag{1}$$

where $a^0 \in \Sigma$ and the sizes of the matrices appearing in the quadruple $(a', \tilde{a}, a^0, 'a)$ make it possible to form the matrix (1). Let

$$^m\mathscr{M}^n(R, \Sigma) = \{a \in \mathscr{M}(R, \Sigma) \mid \tilde{a} \in {}^m R^n\}.$$

Introduce partial addition and multiplication operations on the set $\mathcal{M}(R, \Sigma)$ according to the following rules:

$$a \oplus b = \begin{pmatrix} a' & b' & \tilde{a}+\tilde{b} \\ a^0 & 0 & 'a \\ 0 & b^0 & 'b \end{pmatrix}, \quad a \odot b = \begin{pmatrix} a' & \tilde{a}b' & \tilde{a}\cdot\tilde{b} \\ a^0 & 'ab' & 'a\tilde{b} \\ 0 & b^0 & 'b \end{pmatrix}, \quad (2)$$

assuming that the right-hand sides of these identities are defined if the matrices $\tilde{a}+\tilde{b}$ and $\tilde{a}\cdot\tilde{b}$ are defined, respectively. Also, define a binary relation $\sim$, setting

$$\left(\begin{array}{c|c} a' & \tilde{a} \\ \hline a^0 & 'a \end{array}\right) \sim \left(\begin{array}{c|c} b' & \tilde{b} \\ \hline b^0 & 'b \end{array}\right)$$

if and only if there exists a matrix equality of the form

$$\left(\begin{array}{c|ccc} a' & * & * & * \\ \hline a^0 & * & * & * \\ 0 & c & * & * \end{array}\right) \left(\begin{array}{ccc|c} * & * & * \\ * & * & * \\ \hline d & * & * \\ 0 & b^0 & 'b \end{array}\right) = \left(\begin{array}{cc|c} 0 & -b' & \tilde{a}-\tilde{b} \\ 0 & 0 & 'a \\ 0 & 0 & 0 \end{array}\right), \quad (3)$$

where $c, d \in \Sigma$ and the symbol $*$ stands for certain matrices over R of suitable sizes.

THEOREM 1.1 (V. N. Gerasimov). *Let R be a ring with identity, and let Σ be a multiplicative set of rectangular matrices over R. Then the following statements hold*:

(1) *The binary relation $\sim$ introduced on $\mathcal{M}(R, \Sigma)$ by formula (3) is an equivalence relation. The equivalence class that contains a matrix $a \in \mathcal{M}(R, \Sigma)$ will be denoted as $[a]$ in the sequel.*

(2) *The partial operations $\oplus$ and $\odot$ defined by equalities (2) induce partial addition and multiplication operations on the sets ${}^m\mathcal{M}^n(R, \Sigma)/\sim$ so that ${}^m\mathcal{M}^n(R, \Sigma)/\sim$ can be identified with the set of $(m \times n)$-matrices over the associative ring with identity ${}^1\mathcal{M}^1(R, \Sigma)/\sim$. Under this identification the matrix units are the elements $[(\underline{|}E_{ij}]$, where E_{ij} are standard matrix units in ${}^mR^n$ ($1 \leq i \leq m$, $1 \leq j \leq n$).*

(3) *The ring ${}^1\mathcal{M}^1(R, \Sigma)/\sim$ is isomorphic to $R\Sigma^{-1}$. The universal Σ-inverting homomorphism $\sigma: R \to R\Sigma^{-1}$ can be defined by the rule $\sigma(r) = [(\underline{|}r)]$, and, if $A \in \Sigma$, then $(\sigma(A))^{-1} = [(\frac{-1|0}{A|1})]$; moreover, for every Σ-inverting homomorphism $f: R \to S$, the mapping $\overline{f}: R\Sigma^{-1} \to S$ defined by the formula*

$$\overline{f}\left(\left[\left(\begin{array}{c|c} a' & \tilde{a} \\ \hline a^0 & 'a \end{array}\right)\right]\right) = \tilde{a}^f - (a')^f((a^0)^f)^{-1}('a)^f$$

is a properly defined ring homomorphism that satisfies the condition $\overline{f}\sigma = f$.

Now we state, in the form of a theorem that is a corollary to Theorem 1.1 and equality (3), a criterion of equality to zero for a matrix over $R\Sigma^{-1}$.

THEOREM 1.2 (V. N. Gerasimov). *Let R be a ring with identity, and let Σ be a multiplicative set of rectangular matrices over R. Then the following statements on a matrix $a \in \mathcal{M}(R, \Sigma)$ are equivalent:*

(a) *The element $a \in \mathcal{M}(R, \Sigma)$ represents a zero matrix over the ring $R\Sigma^{-1}$.*

(b) *There exists a matrix equality of the form*

$$\begin{pmatrix} a' & * & * \\ a^0 & * & * \\ 0 & c & * \end{pmatrix} \begin{pmatrix} * & * \\ * & * \\ * & * \\ d & * \end{pmatrix} = \begin{pmatrix} 0 & \tilde{a} \\ 0 & 'a \\ 0 & 0 \end{pmatrix}. \tag{4}$$

(c) *There exists a matrix equality of the form*

$$\begin{pmatrix} * & * & * \\ c_1 & * & * \end{pmatrix} \begin{pmatrix} * & * & * \\ d_1 & * & * \\ 0 & a^0 & 'a \end{pmatrix} = \begin{pmatrix} 0 & a' & \tilde{a} \\ 0 & 0 & 0 \end{pmatrix} \tag{5}$$

(here c, d, c_1 and d_1 are matrices from Σ and the symbol $*$ denotes certain matrices over R of suitable sizes).

COROLLARY. *Let R be a ring with identity, and let Σ be a multiplicative set of rectangular matrices over R. Then the kernel of a universal Σ-inverting homomorphism $\sigma: R \to R\Sigma^{-1}$ consists of those elements $r \in R$ that can be included in a matrix equality of the form*

$$\begin{pmatrix} a' & \tilde{a} \\ a^0 & 'a \end{pmatrix} \begin{pmatrix} b' & \tilde{b} \\ b^0 & 'b \end{pmatrix} = \begin{pmatrix} 0 & r \\ 0 & 0 \end{pmatrix}, \tag{6}$$

where $a^0, b^0 \in \Sigma$ and the remaining matrices over R have suitable sizes. In particular, a set Σ of matrices is potentially invertible if and only if the ring R satisfies all quasi-identities of the form

$$\begin{pmatrix} a' & \tilde{a} \\ a^0 & 'a \end{pmatrix} \begin{pmatrix} b' & \tilde{b} \\ b^0 & 'b \end{pmatrix} = \begin{pmatrix} 0 & r \\ 0 & 0 \end{pmatrix} \Rightarrow r = 0, \tag{7}$$

where $a^0, b^0 \in \Sigma$ and the remaining matrices have suitable sizes.

These statements correspond to Theorem 2 and its corollaries from Gerasimov's paper [13], where their detailed proofs can be found.

§2. Homomorphisms into radical rings

Claims concerning radical rings will be obtained as a particular case of certain more general considerations.

Namely, we consider pairs of rings of the form (R, S), where $R \triangleleft S$. A *pair morphism* $f: (R, S) \to (R_1, S_1)$ is a ring homomorphism $f: S \to S_1$ such that $f(R) \subseteq R_1$.

Fix a ring S with identity, its ideal R, the natural projection $\pi: S \to S/R$, and a multiplicative set Σ of rectangular matrices over S such that $\Sigma \subseteq \pi^{-1}(\mathrm{GL}(S/R))$.

A morphism $f: (R, S) \to (R_1, S_1)$ is called Σ-*inverting* if $f(\Sigma) \subseteq \mathrm{GL}(S_1)$. Among the Σ-inverting morphisms there exists a morphism $\sigma: (R, S) \to (\mathfrak{R}, \mathfrak{S})$ that is universal in the sense that, for every Σ-inverting morphism $f: (R, S) \to (R_1, S_1)$, there exists a uniquely determined morphism $\overline{f}: (\mathfrak{R}, \mathfrak{S}) \to (R_1, S_1)$ that completes the following diagram up to a commutative one:

$$
\begin{array}{ccc}
(R, S) & \xrightarrow{\sigma} & (\mathfrak{R}, \mathfrak{S}) \\
& \searrow f & \downarrow \overline{f} \\
& & (R_1, S_1)
\end{array}
\quad (8)
$$

In the case when $R = S$, the ring S/R is a ring with identity (which equals zero), and hence, we can assume that the restriction $\Sigma \subseteq \pi^{-1}(\mathrm{GL}(S/R))$ is satisfied for any set Σ of matrices over S. Clearly, $\mathfrak{R} = \mathfrak{S} = R\Sigma^{-1}$ in such a case. It turns out that Gerasimov's construction produces a universal Σ-inverting pair in the general case as well. Namely, set $\mathfrak{S} = S\Sigma^{-1}$ and $\mathfrak{R} = \ker \overline{R} = \ker \overline{\pi}$, where $\overline{\pi}: S\Sigma^{-1} \to S/R$ is a homomorphism whose existence is guaranteed by the universality property of the ring $S\Sigma^{-1}$ and the condition $\Sigma \subseteq \pi^{-1}(\mathrm{GL}(S/R)))$, and take the universal Σ-inverting homomorphism $\mathfrak{S} = S\Sigma^{-1}$ for the morphism $\sigma: (R, S) \to (\mathfrak{R}, \mathfrak{S})$. Denote the ideal $\mathfrak{R}$ by the symbol $U(R, \Sigma)$ and prove that the pair of rings $(U(R, \Sigma), S\Sigma^{-1})$ is indeed universal Σ-inverting.

THEOREM 1.3. *Let S be a ring with identity let R be an ideal of S, let $\pi: S \to S/R$ be the natural projection, and let Σ be a multiplicative set of rectangular matrices over S that satisfies the condition $\Sigma \subseteq \pi^{-1}(\mathrm{GL}(S/R))$. Then the following statements hold:*

(1) *The morphism $\sigma: (R, S) \to (U(R, \Sigma), S\Sigma^{-1})$ described Σ-inverting above is universal Σ-inverting.*

(2) *For every matrix $\mathfrak{a}$ over $S\Sigma^{-1}$ there exists an element*

$$a = \left(\begin{array}{c|c} a' & \tilde{a} \\ \hline a^0 & {'a} \end{array} \right) \in \mathscr{M}(S, \Sigma)$$

such that a' and ${'a}$ are matrices over R and $\mathfrak{a} = [a]$. (A matrix a with this property will be called an R-reduced representative for $\mathfrak{a}$.) An element $\mathfrak{a}$ is a matrix over $U(R, \Sigma)$ if and only if, for each of its R-reduced representative a, the entries of the matrix $\tilde{a}$ belong to R. Here the morphisms σ and $\overline{f}$ of diagram (8) can be given by the formulas

$$\sigma(r) = [(\underline{}r)], \quad \overline{f}\left(\left[\left(\begin{array}{c|c} a' & \tilde{a} \\ \hline a^0 & {'a} \end{array}\right)\right]\right) = \tilde{a}^f - (a')^f((a^0)^f)^{-1}({'a})^f.$$

(3) *Two R-reduced elements*

$$a = \left(\begin{array}{c|c} a' & \tilde{a} \\ \hline a^0 & 'a \end{array}\right) \quad \text{and} \quad b = \left(\begin{array}{c|c} b' & \tilde{b} \\ \hline b^0 & 'b \end{array}\right)$$

determine the same matrix over $S\Sigma^{-1}$ *if and only if there exists a matrix equality of the form* (3) *in which* $a^0, b^0, c, d \in \Sigma$ *and the entries of the remaining matrices belong to* R.

PROOF. (1) Clearly, the morphism $\sigma: (R, S) \to (U(R, \Sigma), S\Sigma^{-1})$ is Σ-inverting. To check its universality consider the following commutative diagram

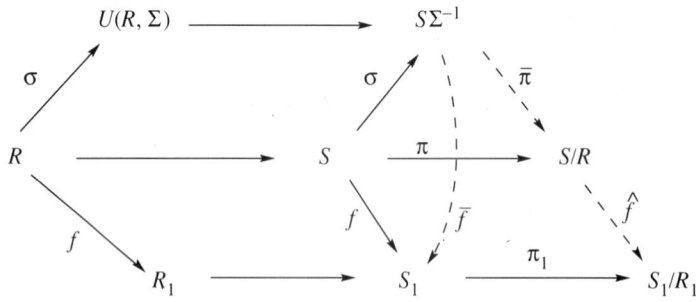

Here $f: (R, S) \to (R_1, S_1)$ is an arbitrary Σ-inverting morphism, π and π_1 are natural projections, the existence of the homomorphisms $\bar{\pi}$ and $\bar{f}$ is guaranteed by the universal property of the ring $S\Sigma^{-1}$, and $\hat{f}(S + R) = f(s) + R_1$. We have to see that the homomorphism $\bar{f}: S\Sigma^{-1} \to S_1$ is a pair morphism i.e., that, $\bar{f}(U(R, \Sigma)) \subseteq R_1$. To this end it suffices to verify the equality $\pi_1 \bar{f}(U(R, \Sigma)) = 0$. It will become obvious if we prove that $\pi_1 \bar{f} = \hat{f}\bar{\pi}$. The latter holds because $\pi_1 \bar{f}\sigma = \pi_1 f = \hat{f}\pi = \hat{f}\bar{\pi}\sigma$ and the homomorphism $\sigma: S \to S\Sigma^{-1}$ is an epimorphism in the category of rings, that is, it is right cancellable.

The remaining properties of the morphism $\bar{f}$ are obvious. This completes the proof of statement (1).

LEMMA 1.1. *Assuming that conditions of Theorem* 1.3 *hold, let* a *and* b *be a pair of matrices over the ring* S *of the form*

$$a = \left(\begin{array}{cc|cc} a'_1 & a'_2 & u'_1 & u'_2 \\ \hline a^0_1 & a_{12} & U_{11} & U_{12} \\ 0 & a^0_2 & U_{21} & U_{22} \end{array}\right), \quad b = \left(\begin{array}{cc|c} V_{11} & V_{12} & 'v_1 \\ V_{21} & V_{22} & 'v_2 \\ \hline b^0_1 & 'b_{12} & 'b_1 \\ 0 & b^0_2 & 'b_2 \end{array}\right),$$

where $a^0_1, a^0_2, b^0_1, b^0_2 \in \Sigma$ *and* a'_1 *and* $'b_2$ *are matrices over* R. *Then there exist matrices* $\alpha, \beta, \gamma, \delta$ *over* S *of the form* $\left(\begin{smallmatrix} 1 & * \\ 0 & 1 \end{smallmatrix}\right)$ *such that the matrices* $\alpha a \beta$ *and* $\gamma b \delta$ *considered as elements of the set* $\mathcal{M}(S, \Sigma)$ *are R-reduced and*

$[\alpha a \beta] = [a]$, $[\gamma b \delta] = [b]$. Moreover, if an equality of the form

$$a \cdot b = r = \left(\begin{array}{cc|c} 0 & r' & \tilde{r} \\ 0 & 0 & 'r \\ 0 & 0 & 0 \end{array}\right), \tag{9}$$

in which r' and $'r$ are matrices over R, holds, then the following conditions hold as well: $\gamma = \beta^{-1}$, $\alpha a \beta \cdot \gamma b \delta = r$

PROOF. Clearly, we can assume at once that (9) holds. Let $c_1^0, c_2^0, d_1^0, d_2^0$ be matrices inverse to $a_1^0, a_2^0, b_1^0, b_2^0$ modulo R, respectively (their existence is secured by the condition $\Sigma \subseteq \pi^{-1}(\mathrm{GL}(S/R))$). Set

$$\alpha = \left(\begin{array}{cc|c} 1 & 0 & -a_2'c_2^0 \\ 0 & 1 & -a_{12}c_2^0 \\ 0 & 0 & 1 \end{array}\right), \quad \delta = \left(\begin{array}{c|c|c} 1 & -\alpha_1^0 b_{12} & -d_1^{0\prime} b_1 \\ 0 & 1 & 0 \\ 0 & 0 & 1 \end{array}\right),$$

$$\beta = \left(\begin{array}{cc|cc} 1 & 0 & -c_1^0 U_{11}' & -c_1^0 U_{12}' \\ 0 & 1 & -c_2^0 U_{21} & -c_2^0 U_{22} \\ 0 & 0 & 1 & 0 \\ 0 & 0 & 0 & 1 \end{array}\right), \quad \gamma = \beta^{-1},$$

where $U_{1i}' = U_{1i} - a_{1i}c_2^0 U_{2i}$ ($i = 1, 2$). We can verify straightforwardly that the matrices $\alpha a \beta$ and $\gamma b \delta$ are R-reduced and satisfy the equality $\alpha a \beta \cdot \gamma b \delta = r$. The equalities $[\alpha a \beta] = [a]$ and $[\gamma b \delta] = [b]$ follow from Lemma 1.2 stated below, which reproduces, for completeness' sake, a fragment of Lemma 4 from Gerasimov's paper [13].

LEMMA 1.2 (V. N. Gerasimov). Let $a = \left(\begin{array}{c|c} a' & \tilde{a} \\ \hline a^0 & 'a \end{array}\right) \in \mathscr{M}(S, \Sigma)$ and let the matrices $\alpha = \left(\begin{array}{c|c} 1 & \alpha' \\ 0 & \alpha^0 \end{array}\right)$ and $\beta = \left(\begin{array}{c|c} \beta^0 & '\beta \\ 0 & 1 \end{array}\right)$ over S be such that $\alpha a \beta \in \mathscr{M}(S, \Sigma)$ and at least one of the matrices α^0 or β^0 is invertible. Then $[a] = [\alpha a \beta]$.

PROOF. We give the proof only in the case when the matrix α^0 is invertible, since the remaining case is considered analogously. Our conclusion follows from the matrix equality

$$\left(\begin{array}{c|c} a' & -\alpha'(\alpha^0)^{-1} \\ \hline a^0 & (\alpha^0)^{-1} \end{array}\right) \left(\begin{array}{c|c} -\beta^0 & -'\beta \\ \hline \alpha^0 a^0 \beta^0 & \alpha^0('a + a^{0\prime}\beta) \end{array}\right)$$
$$= \left(\begin{array}{c|c} -(a' + \alpha' a^0)\beta^0 & -a''\beta - \alpha'('a + a^{0\prime}\beta) \\ \hline 0 & 'a \end{array}\right)$$

and formula (3) of §1.

Lemma 1.2, and hence Lemma 1.1, is proved.

The remaining statements of Theorem 1.3 are deduced from Lemma 1.1.

(2) An R-reduced representative of a matrix $\mathfrak{a} = [(\frac{a' \mid \tilde{a}}{a^0 \mid {}'a})]$ over $S\Sigma^{-1}$ is obtained if we set $a_2^0 = a^0$, $a_2' = a'$, $u_1' = \tilde{a}$, $U_{21} = {}'a$ in Lemma 1.1, while the remaining matrices are assumed to be empty. Furthermore, for a matrix $\mathfrak{a}$ over $S\Sigma^{-1}$ with a fixed R-reduced representative $(\frac{a' \mid \tilde{a}}{a^0 \mid {}'a})$ the following equivalences hold:

($\mathfrak{a}$ is a matrix over $U(R, \Sigma)$)
$$\Leftrightarrow (\overline{\pi}(\mathfrak{a}) = 0) \Leftrightarrow (\pi(\tilde{a}) - \pi(a')(\pi(a^0))^{-1}(\pi({}'a)) = 0)$$
$$\Leftrightarrow (\pi(\tilde{a}) = 0) \Leftrightarrow (\tilde{a} \text{ is a matrix over } R),$$

what was needed.

Statement (3) is an immediate consequence of the definition of equality in the ring $S\Sigma^{-1}$ and Lemma 1.1.

Theorem 1.3 is proved completely.

COROLLARY 1. *Assuming that the conditions of Theorem 1.3 hold, the following statements are equivalent for an R-reduced matrix $a \in \mathcal{M}(S, \Sigma)$:*

(a) *An element $a \in \mathcal{M}(S, \Sigma)$ represents the zero matrix over the ring $S\Sigma^{-1}$.*

(b) *Each of the matrix equalities of the form (4) and (5) holds, where $c, d, c_1, d_1 \in \Sigma$ and the elements of the remaining matrices belong to R.*

COROLLARY 2. *Assuming that the conditions of Theorem 1.3 hold, the kernel of the universal Σ-inverting homomorphsim $\sigma: S \to S\Sigma^{-1}$ consists of those elements $r \in R$ that can be included in a matrix equality of the form (6) with $a^0, b^0 \in \Sigma$, and the elements of the remaining matrices belong to R.*

LEMMA 1.3. *Let S be a ring with identity, and let Σ be a multiplicative set of rectangular matrices over S. If $a = (\frac{a' \mid \tilde{a}}{a^0 \mid {}'a}) \in \mathcal{M}(S, \Sigma) \cap \Sigma$, then the matrix $[a]$ is invertible over the ring $S\Sigma^{-1}$.*

PROOF. We set
$$b = \begin{pmatrix} 0 & 1 & 0 \\ \hline a' & \tilde{a} & -1 \\ a^0 & {}'a & 0 \end{pmatrix} \in \mathcal{M}(S, \Sigma)$$

and verify that $[a] \cdot [b] = [(\underline{1})]$ and $[b] \cdot [a] = [(\underline{1})]$. The first equality follows from the relations below and formula (3). The second equality is verified analogously.

$$a \odot b \ominus (\underline{1}) = \begin{pmatrix} a' & 0 & \tilde{a} & -1 \\ a^0 & 0 & {}'a & 0 \\ 0 & a' & \tilde{a} & -1 \\ 0 & a^0 & {}'a & 0 \end{pmatrix},$$

$$\begin{pmatrix} a' & 0 & \tilde{a} & -1 & 0 \\ a^0 & 0 & 'a & 0 & -1 \\ 0 & a' & \tilde{a} & -1 & 0 \\ 0 & a^0 & 'a & 0 & -1 \end{pmatrix} \begin{pmatrix} 1 & 0 & 0 \\ 1 & 0 & 0 \\ 0 & 1 & 0 \\ a' & \tilde{a} & 1 \\ a^0 & 'a & 0 \end{pmatrix} = \begin{pmatrix} 0 & 0 & -1 \\ 0 & 0 & 0 \\ 0 & 0 & -1 \\ 0 & 0 & 0 \end{pmatrix}.$$

Lemma 1.3 is proved.

Lemma 1.4, which we state below, was proved by Cohn in [33, p. 686] and is given here only for reader's convenience.

LEMMA 1.4 (P. M. Cohn). *Let S be a ring with identity, let R be an ideal in S, let π be the natural projection, and let $\Sigma = \pi^{-1}(\mathrm{GL}(S/R))$. Then $J(S\Sigma^{-1}) = \overline{\pi}^{-1}(\mathrm{GL}(S/R))$, where $\overline{\pi}\colon S\Sigma^{-1} \to S/R$ is the homomorphism whose existence is guaranteed by the universality property of the ring $S\Sigma^{-1}$. In particular, $U(R, \Sigma) = \ker \overline{\pi} \subseteq J(S\Sigma^{-1})$*

PROOF. Since the homomorphism π maps $S\Sigma^{-1}$ onto S/R, the inclusion $J(S\Sigma^{-1}) \subseteq \overline{\pi}^{-1}(J(S/R))$ holds. Conversely, let $\mathfrak{a} = [a]$, $\mathfrak{b} = [b] \in S\Sigma^{-1}$, and $\overline{\pi}(\mathfrak{a}) \in J(S/R)$. Without loss of generality we can assume that the matrices a and b are R-reduced. Then we obtain $1 - ab = [(\underline{1}) \ominus a \odot b]$ and it is not difficult to understand that $(\underline{1}) \ominus a \odot b \in \Sigma$. Lemma 1.3 makes it possible to conclude that the element $1 - ab$ is invertible in the ring $S\Sigma^{-1}$. This means that $\mathfrak{a} \in J(S\Sigma^{-1})$. Lemma 1.4 is proved.

Now we apply the results obtained to investigate the problem of embedding a ring R in a radical ring.

THEOREM 1.4. *Let R be a ring, and let $S = R^{\#}$, let $\pi\colon S \to S/R \cong \mathbb{Z}$ be the natural projection, and let $\Sigma = \pi^{-1}(\mathrm{GL}(S/R))$. Then the following statements hold:*

(1) *The ring $U_J(R) = U(R, \Sigma)$ is radical. It is universal in the sense that, for every homomorphism $f\colon R \to R_1$ into a radical ring R_1, there exists a uniquely determined homomorphism $\overline{f}\colon U_J(R) \to R_1$ that completes the following diagram to a commutative one:*

$$\begin{array}{ccc} R & \xrightarrow{\sigma} & U_J(R) \\ & {\scriptstyle f} \searrow & \downarrow {\scriptstyle \overline{f}} \\ & & R_1 \end{array} \qquad (10)$$

(2) *For every matrix $\mathfrak{a}$ over $S\Sigma^{-1}$ there exists an R-reduced representative $a \in \mathcal{M}(S, \Sigma)$ of the following form $a = \left(\begin{array}{c|c} a' & k+\tilde{a} \\ \hline 1+A & 'a \end{array}\right)$, where $\left(\begin{array}{c|c} a' & \tilde{a} \\ \hline A & 'a \end{array}\right)$ is a matrix over R and k is a matrix over $\mathbb{Z}$. An element $\mathfrak{a}$ is a matrix over $U_J(R)$ if and only if $k = 0$. Here the homomorphisms σ and $\overline{f}$ from diagram (10)*

can be given by the formulas:

$$\sigma(r) = [(\underline{1}r)], \quad \bar{f}\left(\left[\left(\begin{array}{c|c} a' & \tilde{a} \\ \hline 1+A & 'a \end{array}\right)\right]\right) = \tilde{a}^f - (a')^f(1+A^f)^{-1}('a)^f.$$

(3) *Two R-reduced elements* $a = \left(\begin{array}{c|c} a' & \tilde{a} \\ \hline 1+A & 'a \end{array}\right)$ *and* $b = \left(\begin{array}{c|c} b' & \tilde{b} \\ \hline 1+B & 'b \end{array}\right)$ *determine the same matrix over* $U_J(R)$ *if and only if there exists an equality of the form* (3) *in which* $a^0 = 1+A$, $b^0 = 1+B$; $c, d \in 1+R_\infty$, *and the elements of the remaining matrices belong to* R.

PROOF. (1) Since the ring $S/R \cong \mathbb{Z}$ is semisimple, Lemma 1.4 shows that $J(S\Sigma^{-1}) = U_J(R)$; that is, the ring $U_J(R)$ is radical. Clearly, the set Σ consists of square matrices of the form $\lambda + A$, where $\lambda \in GL(\mathbb{Z})$ and A is a matrix over R. Therefore, every homomorphism $f: R \to R_1$ into a radical ring R_1 can be extended to a Σ-inverting pair morphism $f^\#: (R, R^\#) \to (R_1, R_1^\#)$ that acts in accordance with the rule $f^\#(k+r) = k + f(r)$. Now the existence of a homomorphism $\bar{f}: U_J(R) \to R_1$ follows from the universality of the pair $(U_J(R), S\Sigma^{-1})$. The uniqueness of $\bar{f}$ follows from statement (2) of this theorem, since this statement implies that $S\Sigma^{-1} = U_J(R)^\#$.

(2) Let $b = \left(\begin{array}{c|c} a' & k+\tilde{a} \\ \hline b^0 & 'b \end{array}\right)$ be an R-reduced representative of a matrix $\mathfrak{a}$ over $S\Sigma^{-1}$, whose existence was proved in Theorem 1.3. As we have observed above, $b^0 = \lambda + B$, where $\lambda + GL(\mathbb{Z})$ and B is a square matrix over R. Then the matrix $a = \left(\begin{array}{c|c} 1 & 0 \\ \hline 0 & \lambda^{-1} \end{array}\right)\left(\begin{array}{c|c} a' & k+\tilde{a} \\ \hline b^0 & 'b \end{array}\right)$ is, by Lemma 1.2, the desired R-reduced representative.

Analogously, multiplying relation (3) on the left and on the right by suitable invertible matrices, we deduce statement (3) from Theorem 1.3.

Theorem 1.4 is proved.

COROLLARY 1. *Assuming that the conditions of Theorem* 1.4 *hold, the following statements about an R-reduced matrix* $a = \left(\begin{array}{c|c} a' & \tilde{a} \\ \hline 1+A & 'a \end{array}\right) \in \mathcal{M}(S, \Sigma)$ *are equivalent*.

(a) *The element* $\mathfrak{a} \in \mathcal{M}(S, \Sigma)$ *represents a zero matrix over the ring* $S\Sigma^{-1}$.

(b) *Each of the equalities* (4) *and* (5) *holds, where* $c, d, c_1, d_1 \in 1+R_\infty$ *and the elements of the remaining matrices belong to* R.

COROLLARY 2. *The kernel of the universal homomorphism* $\sigma: R^\# \to U_J(R)^\#$ *consists of those elements* $r \in R$ *that can be included in a matrix equality* (6), *where* $a^0, b^0 \in 1+R_\infty$ *and the elements of the remaining matrices belong to* R. *Thus, a ring* R *is embeddable in a radical ring if and only if the set of matrices* $1+R_\infty$ *is potentially invertible. This is true if and only if* R *satisfies all quasi-identities of the form* (7), *in which* $a^0, b^0 \in 1+R_\infty$ *and the elements of the remaining matrices belong to* R.

§3. R-independent sets of matrices

In this section certain sufficient conditions for the injectivity of the universal Σ-inverting morphism $\sigma\colon (R, S) \to (U(R, \Sigma), S\Sigma^{-1})$ will be found. As before, it is assumed that R is an ideal in a ring S, $\pi\colon S \to S/R$ is the natural projection, and $\Sigma \subseteq \pi^{-1}(\mathrm{GL}(S/R))$. In contradistinction to the previous section the set Σ is not assumed to be multiplicative, but it contains the empty (0×0)-matrix.

An arbitrary matrix equality of the form

$$(a|x) \cdot \binom{y}{b} = 0 \tag{11}$$

over S where $a \in \Sigma$ and x is a matrix over R ($b \in \Sigma$ and y is a matrix over R), is said to be a *left (right) matrix (R, Σ)-dependence relation*. If $y = 0$ ($x = 0$), then such a relation of left (right) matrix (R, Σ)-dependence is called *trivial*. An equality of the form (11) is said to be an (R, Σ)-*dependence relation* if $a, b \in \Sigma$ and x and y are matrices over R.

We say that a quadruple $(\alpha, \beta, \gamma, \delta)$ of matrices over S *trivializes* the left (right) matrix (R, Σ)-dependence relation (11) if β and γ are square matrices that admit a partition into blocks

$$\beta = \begin{pmatrix} \beta_{11} & \beta_{12} \\ \beta_{21} & \beta_{22} \end{pmatrix}, \quad \gamma = \begin{pmatrix} \gamma_{11} & \gamma_{12} \\ \gamma_{21} & \gamma_{22} \end{pmatrix},$$

that agree with the partition of the matrices appearing in equality (11), so that $\beta \cdot \gamma = 1$ and the equality

$$\alpha(a|x)\beta \cdot \gamma \binom{y}{b} \delta = 0$$

is a trivial left (right) matrix (R, Σ)-dependence relation.

Following Gerasimov [13] we say that a set Σ of matrices over a ring S is R-*independent* if it satisfies simultaneously the following conditions:

I1. If $ax = 0$ or $xa = 0$ for $a \in \Sigma$ and x a matrix over R, then $x = 0$.

I2. Every matrix (R, Σ)-dependence relation is trivialized by means of a suitable quadruple of matrices over S.

PROPOSITION 1.1. *Let Σ be an R-independent multiplicative set of matrices over a ring S, and let $a = \left(\begin{smallmatrix} a' & \tilde{a} \\ a^0 & 'a \end{smallmatrix}\right)$ be an R-reduced element of $\mathscr{M}(S, \Sigma)$. Then the following statements are equivalent.*

(1) *An element a represents the zero matrix over the ring $S\Sigma^{-1}$.*

(2) *There exists a relation of the form*

$$\begin{pmatrix} a' & \tilde{a} \\ \alpha a^0 & \alpha' a \end{pmatrix} = \begin{pmatrix} b' \\ b^0 \end{pmatrix} \cdot (C^0|'c),$$

where $b^0 \in \Sigma$, b' and c' are matrices over R and c^0, α are matrices over S.

(3) *There exists a relation of the form*

$$\left(\begin{array}{c|c} a'\delta & \tilde{a} \\ \hline a^0\delta & 'a \end{array}\right) = \left(\begin{array}{c} b' \\ \hline b^0 \end{array}\right) \cdot (c^0|'c),$$

where $c^0 \in \Sigma$, b' and c' are matrices over R and b^0, δ are matrices over S.

PROOF. We prove only the equivalence (1) $\Leftrightarrow$ (2), since the equivalence (1) $\Leftrightarrow$ (3) is obtained analogously because of the symmetricity of conditions (2) and (3).

The implication (2) $\Rightarrow$ (1) is true by the equality

$$\left(\begin{array}{c|c} b' & 0 \\ \hline b^0 & \alpha \end{array}\right) \cdot \left(\begin{array}{c|c} -c^0 & -'c \\ \hline a^0 & 'a \end{array}\right) = \left(\begin{array}{c|c} -a' & -\tilde{a} \\ \hline 0 & 0 \end{array}\right)$$

and Theorem 1.2.

Now let $[a] = 0$. By Corollary 1(b) to Theorem 1.3, there exists a matrix equality of the form

$$\left(\begin{array}{cc|c} a' & c_1 & x_1 \\ a^0 & c_2 & x_2 \\ 0 & c^0 & x_3 \end{array}\right) \cdot \left(\begin{array}{c|c} u_1 & u_2 \\ v_1 & v_2 \\ d^0 & d_2 \end{array}\right) = \left(\begin{array}{c|c} 0 & \tilde{a} \\ 0 & 'a \\ 0 & 0 \end{array}\right) \qquad (12)$$

where $a^0, c^0, d^0 \in \Sigma$ and the entries of the remaining matrices belong to R. Let $(\alpha, \beta, \gamma, \delta)$ be a quadruple of matrices that trivializes the (R, Σ)-dependence relation $(c^0|x_3) \cdot \left(\frac{v_1}{d^0}\right) = 0$. Then the equality

$$\left(\begin{array}{ccc} 1 & 0 & 0 \\ 0 & 1 & 0 \\ 0 & 0 & \alpha \end{array}\right) \left(\begin{array}{cc|c} a' & c_1 & x_1 \\ a^0 & c_2 & x_2 \\ 0 & c^0 & x_3 \end{array}\right) \left(\begin{array}{ccc} 1 & 0 & 0 \\ 0 & \beta_{11} & \beta_{12} \\ 0 & \beta_{21} & \beta_{22} \end{array}\right)$$

$$\cdot \left(\begin{array}{ccc} 1 & 0 & 0 \\ 0 & \gamma_{11} & \gamma_{12} \\ 0 & \gamma_{21} & \gamma_{22} \end{array}\right) \left(\begin{array}{c|c} u_1 & u_2 \\ v_1 & v_2 \\ d^0 & d_2 \end{array}\right) \left(\begin{array}{c|c} \delta & 0 \\ \hline 0 & 1 \end{array}\right) = \left(\begin{array}{c|c} 0 & \tilde{a} \\ 0 & 'a \\ 0 & 0 \end{array}\right)$$

is of the form (12) and, additionally, satisfies the conditions $x_3 = 0$ and $v_1 = 0$. However, in this case $v_2 = 0$, since condition I1 holds. Thus, the equality

$$\left(\begin{array}{c|c} a' & x_1 \\ \hline a^0 & x_2 \end{array}\right) \cdot \left(\begin{array}{c|c} u_1 & u_2 \\ \hline d^0 & d_2 \end{array}\right) = \left(\begin{array}{c|c} 0 & \tilde{a} \\ \hline 0 & 'a \end{array}\right)$$

is true. Consider a quadruple $(\alpha, \beta, \gamma, \delta)$ that trivializes the (R, Σ)-dependence relation $(a^0|x_2) \cdot \left(\frac{u_1}{\alpha}\right) = 0$ and set

$$\left(\begin{array}{c|c} 1 & 0 \\ \hline 0 & \alpha \end{array}\right) \left(\begin{array}{c|c} a' & x_1 \\ \hline a^0 & x_2 \end{array}\right) \left(\begin{array}{c|c} \beta_{11} & \beta_{12} \\ \beta_{21} & \beta_{22} \end{array}\right) = \left(\begin{array}{c|c} b' & \tilde{b} \\ \hline b^0 & 0 \end{array}\right),$$

$$\begin{pmatrix} \gamma_{11} & \gamma_{12} \\ \gamma_{21} & \gamma_{22} \end{pmatrix} \begin{pmatrix} u_1 & u_2 \\ d^0 & d_2 \end{pmatrix} = \begin{pmatrix} 0 & \tilde{e} \\ e^0 & 'e \end{pmatrix}.$$

Then $\tilde{b} = 0$, because $\tilde{b}e^0 = 0$ and $e^0 \in \Sigma$. Therefore, we obtain

$$\begin{pmatrix} 1 & 0 \\ 0 & \alpha \end{pmatrix} \begin{pmatrix} a' & \tilde{a} \\ a^0 & 'a \end{pmatrix} = \begin{pmatrix} a' & 0 \\ \alpha a^0 & 0 \end{pmatrix} + \begin{pmatrix} 0 & \tilde{a} \\ 0 & \alpha' a \end{pmatrix}$$

$$= \begin{pmatrix} b' & 0 \\ b^0 & 0 \end{pmatrix} \begin{pmatrix} \gamma_{11} & 0 \\ \gamma_{21} & 0 \end{pmatrix} + \begin{pmatrix} b' & 0 \\ b^0 & 0 \end{pmatrix} \begin{pmatrix} 0 & \tilde{e} \\ 0 & 'e \end{pmatrix}$$

$$= \begin{pmatrix} b' \\ b^0 \end{pmatrix} (\gamma_{11} | \tilde{e}).$$

Proposition 1.1 is proved.

COROLLARY. *Every R-independent multiplicative set Σ of matrices over a ring S is potentially invertible; that is, the morphism $\sigma: (R, S) \to (U(R, \Sigma), S\Sigma^{-1})$ is injective.*

As the following lemma analogous to Lemma 9 of Gerasimov's paper [13] shows, it is not essential that the set Σ be multiplicative in this statement.

LEMMA 1.5. *The multiplicative closure* $\mathrm{mult}(S, \Sigma)$ *of an R-independent set Σ of matrices is R-independent.*

PROOF. We consider an arbitrary (R, Σ)-dependence relation of the form (11) and prove that it can be trivialized by means of a suitable quadruple of matrices over S (the fact that $\mathrm{mult}(S, \Sigma)$ satisfies condition I1 can be verified analogously). We use induction on the sum of the numbers of rows in the matrices a^0 and b^0. If $a^0, b^0 \in \Sigma$ then our relation can be trivialized by this condition. For example, let $a^0 \notin \Sigma$. Then the equality (11) can be written in the form

$$\begin{pmatrix} a_1^0 & a_{12} & x_1 \\ 0 & a_2^0 & x_2 \end{pmatrix} \cdot \begin{pmatrix} y_1 \\ y_2 \\ b^0 \end{pmatrix} = \begin{pmatrix} 0 \\ 0 \end{pmatrix}, \qquad (13)$$

where $a_1^0 \in \Sigma$, a_2^0 is a matrix from $\mathrm{mult}(S, \Sigma)$ whose number of rows is less than that of a^0, x_1, x_2, y_1, y_2 are matrices over R, and a_{12} is a matrix over S.

The $(R, \mathrm{mult}(S, \Sigma))$-dependence relation $(a_2^0 | x_2) \cdot \binom{y_2}{b^0} = 0$ can be trivialized, by the inductive hypothesis, by means of a certain quadruple $(\alpha, \beta, \gamma, \delta)$ of matrices. Let c^0 be the inverse, modulo R, matrix for a_2^0. We can verify straightforwardly that the quadruple

$$\left(\begin{pmatrix} 1 & -a_{12}c^0 \\ 0 & \alpha \end{pmatrix}, \begin{pmatrix} 1 & 0 \\ 0 & \beta \end{pmatrix}, \begin{pmatrix} 1 & 0 \\ 0 & \gamma \end{pmatrix}, \delta \right)$$

of matrices transforms (13) into an $(R, \text{mult}(S, \Sigma)$-dependence relation of the same form which satisfies additional conditions $x_2 = 0$ and $y_2 = 0$. It remains to apply the induction hypothesis once more to the relation $(a_1^0|x_1) \cdot (\frac{y_1}{b_1^0}) = 0$.

Lemma 1.5 is proved.

Now define a certain class of pairs of rings and a set $\Sigma \subseteq \pi^{-1}(\text{GL}(S/R))$ of matrices for each pair (R, S) from this class in such a way that the set Σ is R-independent.

As a preliminary, we introduce necessary notions and definitions.

A $(k \times k)$-matrix formed by the entries appearing at the intersection of certain k rows and k columns of a matrix $A \in {}^n S^m$ ($k \leq \min\{m, n\}$) is said to be a *minor of order k* of A.

Let Σ be a set of square matrices over a ring S. An element $A \in {}^n S^m$ is called a Σ-*matrix* if there exists a minor of order $\min\{m, n\}$ of A that belongs to Σ. The set of Σ-matrices of size $(m \times n)$ is denoted by the symbol ${}_m\Sigma_n$, and if $m = n$, we write simply Σ_m. The symbol $\text{r} - \text{GL}_m(S)$ $(1 - \text{GL}_m(S))$ denotes the set of the right (left) invertible square matrices of order m over the ring S. Let $\text{GL}_m(S) = (\text{r} - \text{GL}_m(S)) \cap (1 - \text{GL}_m(S))$.

Let S be a ring with identity and R its ideal, let $\pi: S \to S/R$ be the natural projection, and let $\mathscr{T} \subseteq \pi^{-1}(\text{GL}(S/R)) \cap S$. A set $\Sigma \subseteq \pi^{-1}(\text{GL}(S/R))$ of square matrices is called an *n-proper extension* of the set $\mathscr{T}$ if the following conditions hold simultaneously:

P1. For each matrix $a \in \Sigma_m$ ($m \leq n$) there exist matrices $P, Q \in \text{GL}_m(S)$ such that $PaQ = \begin{pmatrix} \tilde{a} & a' \\ a & A \end{pmatrix}$, $\tilde{a} \in \mathscr{T}$.

P2. The operation of striking out the ith column (the ith row) turns the set Σ_m into a set of Σ-matrices (for $m \leq n$).

P3. The following equalities hold: $({}_m\Sigma_k) \cdot (\text{r} - \text{GL}_k(S)) = {}_m\Sigma_k$ for $m \leq k \leq 2n$ and $(1 - \text{GL}_m(S)) \cdot ({}_m\Sigma_k) = {}_m\Sigma_k$ for $k \leq m \leq 2n$.

A ring S is called $(n, \mathscr{T})$-*free over R* if every left (right) $(R, \mathscr{T})$-dependence relation of length not exceeding n can be trivialized by means of a suitable quadruple of square matrices over S of the form $(1, \beta, \gamma, 1)$. In such a case we say that (β, γ) is a trivializing pair of matrices. Recall that the length of relation (11) is the number of columns in the matrix $(a|x)$.

THEOREM 1.5. *Let a ring S be a $(2n, \mathscr{T})$-free over R, and let Σ be an arbitrary n-proper extension of the set $\mathscr{T}$. Then, for every $m \leq n$, the set Σ_m is R-independent.*

PROOF. The claim of the theorem will be deduced from the following lemma.

LEMMA 1.6. *Assuming that the conditions of Theorem 1.5 hold, every left (right) (R, Σ_m)-dependence relation of length not exceeding $2n$ can be trivialized by means of a suitable pair of matrices.*

PROOF. The proof is by induction on the number m, restricting ourselves to a verification of the fact that left (R, Σ_m)-dependence relations can be trivialized. The case $m = 1$ holds because the ring S is $(2n, \mathscr{T})$-free over R and conditions P1 and P3 hold.

Now consider an arbitrary left (R, Σ_m)-dependence relation $(m > 1)$: $(a|x) \cdot \binom{y}{b} = 0$, where $a \in \Sigma_m$, x is a matrix over R, and the length of the relation does not exceed $2n$. Using condition P1 and acting on the original relation by a quadruple $(P, (\frac{Q|0}{0|1}), (\frac{Q^{-1}|0}{0|1}), 1)$ of matrices, where P and Q are invertible matrices, whose existence is guaranteed by condition P1, we arrive at an equality of the form

$$u \cdot v \equiv \begin{pmatrix} \tilde{a} & a' & x' \\ 'a & A & X \end{pmatrix} \cdot \begin{pmatrix} \tilde{y} \\ 'y \\ \overline{b} \end{pmatrix} = 0 \tag{14}$$

in which $\tilde{a} \in \mathscr{T}$. Since $\mathscr{T} \subseteq \pi^{-1}(\mathrm{GL}(S/R))$, we can assume that a' and $'a$ are matrices over R. The left $(R, \mathscr{T})$-dependence relation $(\tilde{a}|a'x') \cdot \binom{\tilde{y}}{y}{b} = 0$ can be trivialized by our condition. Therefore, we can assume from the beginning that $\tilde{y} = 0$. Then the equality (14) need not be a left (R, Σ_m)-dependence relation. Nevertheless, condition P3 allows us to assume that the left factor in this equality is a Σ_m-matrix. If a Σ_m-minor exists that does not contain the first column of the matrix u, then permuting the columns of u, we arrive at an equality of the form $(a|x) \cdot \binom{y}{b} = 0$ whose length is less than that of the original left (R, Σ_m)-dependence relation. The equality we obtain satisfies the condition $a \in \Sigma_m$, although the entries of the matrix x do not have to belong to R. However, this can be easily amended acting on this relation by a pair of matrices of the form $((\frac{1|-cx}{0|1}), (\frac{1|cx}{0|1}))$, where c is a matrix inverse to a modulo R. Thus, in this case the proof is completed by induction on the length of the relation.

Now assume that all Σ_m-minors of the matrix u contain its first column. An argument analogous to that given above yields a relation of the form (14) in which $\tilde{y} = 0$ and $a = \begin{pmatrix} \tilde{a} & a' \\ 'a & A \end{pmatrix} \in \Sigma_m$. By condition P2, the matrix $\binom{a'}{A}$ contains a Σ_{m-1} minor. Permuting, if necessary, rows in the matrix u, we can assume that $A \in \Sigma_{m-1}$ (and then $\tilde{a}$ is not necessarily in $\mathscr{T}$ any more). Applying now the inductive hypothesis to the left (R, Σ_{m-1})-dependence relation $(A|X) \cdot \binom{'y}{b} = 0$ we arrive at an equality of the form (14) that satisfies the conditions $\tilde{y} = 0$ and $'y = 0$. The matrix u in this equality remains an Σ_m-matrix. If $a = \begin{pmatrix} \tilde{a} & a' \\ 'a & A \end{pmatrix} \in \Sigma_m$, then everything is proved, and if $a \notin \Sigma_m$, then the argument used during our consideration of the previous case make it possible to decrease the length of the left (R, Σ_m)-dependence relation.

Observe that multiplying left (R, Σ_m)-dependence relations on the left by invertible matrices that we used in our argument does not affect the fact that these relations can be trivialized by means of a pair of matrices (by condition P3).

Lemma 1.6 is proved.

Now we return to the proof of Theorem 1.5. Condition I1 follows from Lemma 1.6 straightforwardly.

To verify condition I2 observe that every (R, Σ_m)-dependence relation can be considered as a left (R, Σ_m)-dependence relation. Since these latter relations can be trivialized by Lemma 1.6, it suffices to understand that for every matrix of the form $(\frac{y}{b})$, where $b \in \Sigma_m$, and a matrix $\gamma = \left(\begin{array}{c|c}\gamma_{11} & \gamma_{12} \\ \hline \gamma_{21} & \gamma_{22}\end{array}\right) \in$ $1 - \mathrm{GL}(S)$, if $\left(\begin{array}{c|c}\gamma_{11} & \gamma_{12} \\ \hline \gamma_{21} & \gamma_{22}\end{array}\right)(\frac{y}{b}) = (\frac{0}{c})$ then $c \in \Sigma_m$. Indeed, by condition P3 the matrix $(\frac{0}{c})$ contains a Σ_m-minor. Therefore, it remains to understand that a matrix from Σ_m cannot contain zero rows for $m \leq n$ (if only $S \neq \{0\}$). This holds if $m = 1$, for otherwise, by condition P1, we obtain $0 \in \mathscr{T}$ which contradicts the fact that the $\mathscr{T}$-dependence relation $0 \cdot x = 0$ can be trivialized, the condition $\mathscr{T} \subseteq \pi^{-1}(\mathrm{GL}(S/R))$, and the existence of an identity element (different from zero) in the ring S. In the general case it is easy to make the induction step using conditions P1 and P2.

Theorem 1.5 is proved.

We note the most important particular cases of Theorem 1.5. Let $R = S$, $\mathscr{T} = S \setminus \{0\}$. Then the definition of the $(n, \mathscr{T})$-freedom of S over R becomes the definition of an n-FI-ring (see Cohn's book [16, §1.1]). Besides, using the simplest properties of full matrices over $2n$-FI-rings (contained in the book we have just referred to), it is easy to understand that the set Σ of all full matrices over an $2n$-FI-ring S is an n-proper extension of the set $\mathscr{T} = S \setminus \{0\}$. Thus, in this case Theorem 1.5 produces a known statement on the independence of the set of full $(n \times n)$-matrices over an $2n$-FI-ring.

Another particular case touches directly upon embeddings in radical rings. Namely, let R be a ring let $S = R^{\#}$, and let $\mathscr{T} = 1 + R$. Obviously, the set $\Sigma = \pi^{-1}(\mathrm{GL}(S/R))$ of matrices is an n-proper extension of the set $\mathscr{T}$ for every n. In this case the rings R for which the ring $S = R^{\#}$ is $(n, \mathscr{T})$-free over R are called A_n-rings. Thus, a ring R is an A_n-ring if it satisfies simultaneously the following two conditions:

(α_n). For every number $m \leq n$ and an arbitrary equality of the form

$$(1 + x|x') \cdot \left(\frac{y}{'y}\right) = 0,$$

where $x \in R$, $x' \in R^{m-1}$ and y and $'y$ are matrices over $R^{\#}$, there exists a quasi-invertible matrix $T \in R_m$ such that $(1 + T)^{-1}(\frac{y}{'y}) = (\frac{0}{*})$.

$(\alpha_n)'$. For every number $m \leq n$ and an arbitrary equality of the form $(x'|x) \cdot (\frac{'y}{1+y}) = 0$ where $y \in R$, $'y \in {}^{m-1}R$, and x and x' are matrices over $R^{\#}$, there exists a quasi-invertible matrix $T \in R_m$ such that $(x'|x)(1 + T) = (*|0)$.

Theorem 1.5, Proposition 1.1, and Lemma 1.5 imply the following

PROPOSITION 1.2. *Let R be an A_{2n}-ring. Then the set $1 + R_n$ of matrices over $R^{\#}$ is potentially invertible. In particular, a ring that is an A_n-ring for all natural n is embeddable in a radical ring.*

The following lemma clarifies a connection between A_n-rings and n-FI-rings.

LEMMA 1.7. *Let $R^{\#}$ be an n-FI-ring for a certain n. Then R is an A_n-ring.*

PROOF. First we check that the ring R satisfies condition (α_n) (condition $(\alpha_n)'$ is verified analogously).

Consider an arbitrary equality of the form $(1 + x|x') \cdot (\frac{y}{'y}) = 0$, where $x \in R$, $x' \in R^{m-1}$ and y and $'y$ are matrices over $R^{\#}$. Let P be an invertible matrix of order m over $R^{\#}$ such that the nonzero components of the vector $(1 + x|x') \cdot P$ are right-independent over $R^{\#}$ (the existence of such a matrix follows from the fact that $R^{\#}$ is an n-FI-ring). Clearly, we can assume that the equality $(1 + x|x') \cdot P = (1 + z|z')$, where $z \in R$, $z' \in R^{m-1}$ holds. Then $P = (\frac{1+\tilde{P} \mid P'}{'\lambda +' P \mid \lambda + P^0})$, where $(\frac{1 \mid 0}{'\lambda \mid \lambda}) \in \mathrm{GL}_m(\mathbb{Z})$ and $(\frac{\tilde{P} \mid P'}{'P \mid P^0})$ is a matrix over R. Also, $P \cdot (\frac{y}{'y}) = (\frac{0}{*})$, and hence, we can take the matrix $(\frac{1 \mid 0}{\lambda^{-1/\lambda} \mid \lambda^{-1}})(\frac{\tilde{P} \mid P'}{'P \mid P^0})$ as the desired quasi-invertible matrix T.

Lemma 1.7 is proved.

§4. Examples of A_n-rings

All the examples given below are based on Corollary 2 to Theorem 1.4, Proposition 1.2, Lemma 1.7, and the following theorem of Cohn (see [29, Theorem 3.1]) which is formulated here for the reader's sake (in a somewhat simplified version).

THEOREM 1.6 (Cohn). *Let H be an algebra with identity over a field F presented by a set Z of generators and the defining relations*

$$\sum_{k=1}^{n+1} x_{ik} y_{kj} = l_{ij},$$

where x_{ik}, y_{kj} ($i \in I$, $j \in J$, $k = 1, \ldots, n+1$) are arbitrary pairwise distinct elements of Z and l_{ij} are certain F-linear combinations of elements of Z. Then H is an n-FI-ring.

Obviously, if the conditions of the Cohn theorem hold, then $\mathscr{H} = F \dotplus R$, where R is the ideal generated by the set Z.

EXAMPLE 1. Let H be an algebra with identity over a field F presented by a set $Z = \{x_{ij}, y_{ij} | 1 \leq i, j \leq n+1\}$ of generators and the defining

relations
$$\sum_{k=1}^{n+1} x_{ik} y_{kj} = -y_{ij} \qquad (1 \leq i, j \leq n+1).$$

By Cohn's theorem, the algebra H is an n-FI-ring. By Lemma 1.7, the corresponding ring R is an A_n-ring. However, the matrix equality $(1+X) \cdot Y = 0$ holds over R, where X and Y are matrices formed by the equivalence classes of the elements x_{ij} and y_{ij} ($1 \leq i, j \leq n+1$), respectively. This shows that R is not embeddable in a radical ring (in fact, R is not an A_{n+1}-ring.

Thus, we have proved the fundamental

THEOREM 1.7. *The quasivariety of rings embeddable in radical rings cannot be defined by a finite system of quasi-identities.*

EXAMPLE 2 (Andrunakievich [1]). Let H be an algebra with identity over a field F presented by a set $Z = \{x_{ij}, y_{ij} | 1 \leq i, j \leq 2\}$ of generators and defining relations
$$\sum_{k=1}^{2} x_{ik} y_{kj} = x_{i2} - y_{2j} - x_{i1} - y_{1j}, \qquad (i, j) \neq (2, 2).$$

By Cohn's theorem, the ring H is an integral domain. Therefore, R is an A_1-ring, i.e., its adjoined semigroup satisfies the two-sided cancellativity conditions. However, it is not embeddable in a group because it satisfies the equalities $x_{i2} \circ (-y_{2j}) = x_{i1} \circ y_{1j}$, where $(i, j) \neq (2, 2)$, and $x_{22} \circ (-y_{22}) \neq x_{21} \circ (y_{12})$, which contradicts Mal'tsev's conditions [20] for embeddability of semigroups in groups.

In conclusion, consider the following classes of rings: $\mathscr{P}_0$ of A_1-rings, $\mathscr{P}_1$ of rings whose adjoined semigroup is embeddable in a group, $\mathscr{P}_2$ of potentially quasi-invertible rings, and $\mathscr{P}_3$ of rings embeddable in radical rings. Clearly, the following inclusions take place:
$$\mathscr{P}_0 \supseteq \mathscr{P}_1 \supseteq \mathscr{P}_2 \supseteq \mathscr{P}_3.$$

The example of Andrunakievich shows that $\mathscr{P}_0 \neq \mathscr{P}_1$. The inequality $\mathscr{P}_2 \neq \mathscr{P}_3$ follows from the existence of A_n-rings ($n \geq 2$) that are not embeddable in radical rings (see Example 1). An example showing that $\mathscr{P}_1 \neq \mathscr{P}_2$ will be produced in the next chapter of this dissertation.

We remark that it is not known at present whether quasivarieties of rings embeddable in radical rings and direct products of division rings, respectively, could be defined by systems of quasi-identities that depend on a finite number of variables. Also, it is not clear whether there exist systems of independent quasi-identities defining these quasivarieties.

Chapter 2
Examples of Noninvertible Rings Embeddable in Groups

Recall that an associative ring is called *invertible* (*quasi-invertible*) if each of its nonzero elements is invertible (quasi-invertible) in a certain larger ring. The quasi-invertibility is understood as the invertibility with respect to the operation of adjoined multiplication $a \circ b = a + b + ab$. In the case when the multiplicative semigroup of a ring is embeddable in a group, we say that this ring is embeddable in a group.

The problem of existence of rings that satisfy the property formulated in the title appeared in connection with the problem of Mal′tsev (see [21, p. 6]) concerning the example of a ring that is not embeddable in a division ring but is embeddable in a group. This problem has been solved independently by Bowtell [27], Klein [37], and Bokut′ [4]–[6] who constructed an algebra over a field of two elements that is embeddable in a group but not invertible. The problem on the existence of the similar examples for fields of arbitrary characteristic was formulated in [15] (Problem 2.22).

In this chapter we prove that the desired property is enjoyed by a series of rings introduced by Gerasimov [13] while looking for conditions necessary and sufficient for invertibility of arbitrary associative rings.

Namely, let $\left(\frac{a' \mid \tilde{a}}{a^0 \mid {}'a}\right)$ and $\left(\frac{b' \mid \tilde{b}}{b^0 \mid {}'b}\right)$ be block decompositions of two general matrices whose product is defined and can be found along the multiplication rules for block matrices. Also, assume that a^0 and b^0 are square upper-triangular matrices, and that the order of each of them is no less than two. Then the algebra with identity over an arbitrary field determined by the following relations written in the matrix form:

$$\left(\frac{a' \mid \tilde{a}}{a^0 \mid {}'a}\right) \cdot \left(\frac{b' \mid \tilde{b}}{b^0 \mid {}'b}\right) = \left(\frac{0 \mid *}{0 \mid 0}\right) \tag{1}$$

is embeddable in a group but not invertible. (Here and in the sequel an asterisk denotes elements whose values are not restricted.) A proof of this statement is the principal goal of this chapter.

Problems analogous to those formulated above also appear when studying embeddings of rings in radical rings. Up to now not a single example of a ring that is not quasi-invertible but whose adjoined semigroup is embeddable in a group was known.

We will prove that such are algebras without identity presented by defining relations of the form

$$\left(\begin{array}{c|c} a' & \tilde{a} \\ \hline 1+a^0 & 'a \end{array}\right) \left(\begin{array}{c|c} b' & \tilde{b} \\ \hline 1+b^0 & 'b \end{array}\right) = \left(\begin{array}{c|c} 0 & * \\ \hline 0 & 0 \end{array}\right), \qquad (2)$$

where the identity is used purely formally and both $\left(\begin{smallmatrix} a' & \tilde{a} \\ a^0 & 'a \end{smallmatrix}\right)$ and $\left(\begin{smallmatrix} b' & \tilde{b} \\ b^0 & 'b \end{smallmatrix}\right)$ are matrices of the above described structure.

In §1 of this chapter we study the structure of the multiplicative semigroup of the ring (1) in order to complete the ring. Here the semigroup considered can be described in terms of generators and relations analogously to the description of multiplicative semigroups of $\overline{SN}$-rings introduced by Bokut' [3] in his solution of Mal'tsev problem. Since we do not need such a description in full, we only give a proof of a theorem that is equivalent to Theorem 2 of [3.II]. The argument itself is a simplified version of the proofs suggested by Bokut' [3.II] and is based on the notion of trivializability of relations introduced in §3 of Chapter I of the dissertation.

In §2 we show that the multiplicative (adjoined) semigroup of the ring (1) (the ring (2), respectively) is embeddable in a group. The idea of the proof is based on a method of Koshevoĭ [17] of embedding an algebra with a single defining relation $Wh = uf$ in a free associative algebra.

§1. The structure of the semigroup $\overline{R^*}$

1. Fix an arbitrary field $\mathscr{F}$, and let R denote an algebra with identity over $\mathscr{F}$ defined by the relations (1) rewritten in the form

$$\left(\begin{array}{cc|cc} V & 'v & 't & T \\ W & 'w & 'u & U \\ 0 & w & u & u' \end{array}\right) \left(\begin{array}{cc|c} 'h & H & G \\ h & h' & g' \\ f & f' & e' \\ 0 & F & E \end{array}\right) = \left(\begin{array}{cc|c} 0 & 0 & * \\ 0 & 0 & 0 \\ 0 & 0 & 0 \end{array}\right), \qquad (3)$$

where V, T, W, U, H, G, F, E are general matrices of sizes $k \times l$, $k \times m$, $l \times l$, $l \times m$, $l \times m$, $l \times n$, $m \times m$, $m \times n$, respectively, and such that the matrices W and F are upper-triangular, $'v$, $'t$ and $'w$, $'u$, $'h$ (g', e' and h', f', u') are columns (rows) of length k and l (n and m), respectively, and w, u, h, f are letters. Here k, l, m, n are natural numbers that satisfy the condition $l + m \geq 2$.

We stipulate that the entries of the general matrices singled out above will be endowed with two indices in calculations, while the entries of rows and the columns (except u' and $'h$) with a single index. Elements $w(h)$, $u(f)$, and the ith component of the vector $u'('h)$ will be denoted by $w_{l+1}(h_0)$, $u_{l+1}(f_0)$, and $u_{l+1\,i}(h_{i0})$, respectively.

Fix an arbitrary ordering of the letters occurring in the relations (3) in such a way that each of the elements v_i $(1 \le i \le k)$, w_j $(1 \le j \le l+1)$ be the greatest in its row. This ordering can be naturally extended to the free semigroup with the letters from (3) as free generators. Here the words

$$\begin{aligned} v_i h_j & \quad (1 \le i \le k,\ 0 \le j \le m), \\ w_i h_j & \quad (1 \le i \le l+1,\ 0 \le j \le m), \\ w_i g_j & \quad (1 \le i \le l+1,\ 1 \le j \le n), \end{aligned} \qquad (4)$$

are the senior words in the relations (3) and the set of relations itself is closed under compositions. By Shirshov's lemma (see [2.1, p. 53]), a basis of the algebra R is formed by all words that do not contain subwords of the form (4). We call such words *basic*. In particular, we obtain that no entry of the matrix $*$ equals zero in the algebra R.

Observe that the algebra R is not invertible. Indeed, if all of its nonzero elements are invertible in a certain larger ring, then the upper-triangular matrices a^0 and b^0 from (1) would have inverses over that ring and then

$$* = a'\tilde{b} + \tilde{a}'b = -a'((a^0)^{-1})'a + b'(b^0)^{-1})'b = 0$$

which contradicts the above remark on the matrix $*$. It will be proved in §2 that the multiplicative semigroup of the ring R is embeddable in a group; thus, R is the desired example of a noninvertible ring that is embeddable in a group.

Before we move to the main results of this section we make a few additional simple remarks.

The function of lower degree $d: R \to \mathbb{N} \cup \{+\infty\}$ is defined in the standard way; it takes value 1 on each of the letters occurring in (1) and satisfies the conditions:

(1) $d(A) = +\infty \Leftrightarrow A = 0$,
(2) $d(A + B) \ge \min\{d(A), d(B)\}$,
(3) $d(A \cdot B) \ge d(A) + d(B)$ for any elements $A, B \in R$.

With its help we introduce a norm on R by the formula $\|A\| = e^{-d(A)}$. The completion of R with respect to this norm is denoted by $\overline{R}$. Thus, $\overline{R}$ is the set of formal power series of the form

$$A = \sum_{i=0}^{\infty} A_i, \qquad (5)$$

where the symbol A_i denotes a homogeneous of degree i element of the ring R with the ordinary operations of addition and multiplication.

Let $p(x_1, \ldots, x_s)$ be a noncommutative polynomial, and let X_i $(1 \le i \le s)$ be the elements of $\overline{R}$ represented in the form of linear combinations of basic words. If subwords of the form (4) do not appear when the element $p(X_1, \ldots, X_s)$ is reduced to the same form, this element is called *reduced*.

Let $A = \sum_{i=0}^{\infty} A_i \in \overline{R}$, and let σ be a basic word of a positive degree l. Represent the element $\sum_{i=l}^{\infty} A_i$ as a formal right (left) linear combination of basic words of length l with coefficients from $\overline{R}$: $\sum_k \sigma_k B_k$ ($\sum_k B_k \sigma_k$), where every product $\sigma_k B_k$ ($B_k \sigma_k$) is reduced. Let ${}^\sigma\!A$ (A^σ) denote the coefficient of σ in such a presentation. If ${}^\sigma\!A = 0$ ($A^\sigma = 0$), then we say that the element A does not begin with σ (does not end with σ). Sometimes it is useful to define the symbols ${}^\sigma\!A$ (A^σ) also for a nonzero element $\sigma \in \mathscr{F}$ by setting them equal to $\sigma^{-1} A$.

From the form of the relations (4) we directly obtain that for any elements $A, B, C \in \overline{R}$ ($d(B) > 0$), if the product $A \cdot B$ is reduced, then the element $A \cdot (B \cdot C)$ is reduced as well, provided $B \cdot C$ is written in the form of a linear combination of basic words.

2. Recall that each equality of the form $(A|C) \cdot (\frac{B}{D}) = 0$, where $A, B, C, D \in \overline{R}$, is called a dependence relation. In what follows when considering dependence relations, we assume that at least one of the four participating elements differs from zero. If $\alpha, \delta \in \overline{R}^*$ and β and γ are mutually inverse matrices of the second order over $\overline{R}$, we say that the dependence relation $\alpha(A|C)\beta \cdot \gamma(\frac{B}{D})\delta = 0$ is obtained from the relation $(A|C) \cdot (\frac{B}{D}) = 0$ by the action of the quadruple $(\alpha, \beta, \gamma, \delta)$ of matrices.

The following theorem is the principal instrument which helps us to obtain necessary information on the structure of the semigroup $\overline{R}^*$. It is equivalent to Theorem 2 of [3.II].

THEOREM 2.1. *Every dependence relation*

$$(A|C) \cdot \left(\frac{B}{D}\right) = 0, \qquad (6)$$

where $A, B, C, D \in \overline{R}$, can be obtained by the action of a suitable quadruple $(\alpha, \beta, \gamma, \delta)$ of matrices on one of the dependence relations of the following form:
   (1) $(1|0) \cdot (\frac{0}{1}) = 0$,
   (2) $(w|u) \cdot (\frac{h}{j}) = 0$.

First we prove this theorem for homogeneous elements A, B, C, and D; then we switch to the general case. Here we use a simplified version of an argument suggested by Bokut' in [3.II].

PROPOSITION 2.1. *Every homogeneous dependence relation of the form* (6) *in which $d(C) \leq d(A)$ has one of the following forms*:
   (a)

$$A = CX, \qquad D = -XB,$$

$$\left((\alpha, \beta, \gamma, \delta) = \left(c, \left(\begin{array}{c|c} X & 1 \\ \hline 1 & 0 \end{array}\right), \left(\begin{array}{c|c} 0 & 1 \\ \hline 1 & -X \end{array}\right), B\right)\right), \qquad (7)$$

(b)
$$A = X(w + u\beta_{21})\lambda, \qquad B = \mu(h - \beta_{12}f)Y,$$
$$C = X(u + w\beta_{12})\xi, \qquad D = \eta(f - \beta_{21}h)Y,$$
$$((\alpha, \beta, \gamma, \delta) = \left(X, \left(\begin{array}{c|c}\lambda & \beta_{12}\xi \\ \hline \beta_{21}\lambda & \xi\end{array}\right), \left(\begin{array}{c|c}\mu & -\beta_{12}\mu \\ \hline -\eta\beta_{21} & \eta\end{array}\right), Y\right)), \qquad (8)$$

where X, Y are homogeneous elements of R, $\beta_{ij}, \lambda, \mu, \xi, \eta \in \mathscr{F}$, and $\Delta = 1 - \beta_{21}\beta_{12} \neq 0$, $\lambda\mu = \xi\eta = \Delta^{-1}$.

(c) *The quadruple (A, B, C, D) is obtained from the quadruple of the form described in* (b) *by means of the operation*
$$P(A, B, C, D) = (C, D, A, B).$$

PROOF. Write the elements C and D in the form
$$C = \sum_{i=1}^{k} C_{1i}v_i + \sum_{i=1}^{l+1} C_{2i}w_i + C_0,$$
$$D = \sum_{j=1}^{n} g_j D_{1j} + \sum_{j=0}^{m} h_j D_{2j} + D_0, \qquad (9)$$

where C_0 does not end with v_i $(1 \leq i \leq k)$, w_j $(0 \leq j \leq l)$, and D_0 does not begin with g_i $(1 \leq i \leq n)$, h_j $(0 \leq j \leq m)$. Then

$$C \cdot D = \sum_{i=1}^{k}\sum_{j=1}^{n} C_{1i} v_i g_j D_{1j}$$
$$- \sum_{i=1}^{k}\sum_{j=0}^{m} C_{1i} \left(\sum_{s=1}^{l} v_{is}h_{sj} + t_i f_j + \sum_{s=1}^{m} t_{is}f_{sj}\right) D_{2j}$$
$$- \sum_{i=1}^{l+1}\sum_{j=1}^{n} C_{2i} \left(\sum_{s=1}^{l} w_{is}g_{sj} + u_i e_j + \sum_{s=1}^{m} u_{is}e_{sj}\right) D_{1j} \qquad (10)$$
$$- \sum_{i=1}^{l+1}\sum_{j=0}^{m} C_{2i} \left(\sum_{s=1}^{l} w_{is}h_{sj} + u_i f_j + \sum_{s=1}^{m} u_{is}f_{sj}\right) D_{2j}$$
$$+ C_0 D + \sum_{i=1}^{k} C_{1i} v_i D_0 + \sum_{i=1}^{l+1} C_{2i} w_i D_0$$

is a reduced presentation of the element $C \cdot D$ (here $f_{sj} = 0$ for $s > j$ and $w_{is} = 0$ for $i > s$). The proof is by induction on the degree of the element C. First let $d(C) = 1$. Then $C_{1i}, C_{2j} \in \mathscr{F}$ in the representation (9).

Consider several possible cases.

I. $d(A) > 1$.

Consider consecutively two possibilities.

I.1. $D_0 = 0$. Comparing the terms in (6) that begin with $v_i g_j$ $(1 \leq i \leq k, 1 \leq j \leq n)$ and $t_i f_j$ $(1 \leq i \leq k, 0 \leq j \leq m)$, we arrive at the equalities

$C_{1i}D_{1j} = -({}^{v_i g_j}A)B$, $C_{1i}D_{2j} = ({}^{t_i f_j}A)B$, which imply that if at least one of C_{1i} $(1 \le i \le k)$ differs from zero, then $D = -XB$ for some $X \in R$. In exactly the same way, if at least one of the C_{2i} $(1 \le i \le l+1)$ differs from zero, then $D = -XB$ for some $X \in R$. In exactly the same way, if at least one of the C_{2i} $(1 \le i \le l+1)$ is different from zero, then the equality $D = -XB$ is obtained by comparing in (6) the terms beginning with $u_i e_j$ $(1 \le i \le l+1, 1 \le j \le n)$ and $u_i f_j$ $(1 \le i \le l+1, 0 \le j \le m)$. Therefore, we can assume that $C_{1i} = C_{2j} = 0$ $(1 \le i \le k, 1 \le j \le l+1)$. Then an analogous result is obtained by comparing in (6) the terms beginning with any letter that occurs in the representation (9) of the element C with a nonzero coefficient. Thus, we obtain case (a) of Proposition 2.1.

We note in passing that the absence of zero divisors in R (and hence in $\overline{R}$) can be deduced from Theorem 1.6 of Cohn or obtained directly by methods similar to those we use. Therefore, in the sequel this property of the ring $\overline{R}$ will be used without special mention.

I.2. $D_0 \ne 0$. Comparing in (6) terms beginning with v_i but not with $v_i g_j$ $(1 \le i \le k, 1 \le j \le n)$, and with w_i $(1 \le i \le l+1)$, we obtain (provided that at least one of C_{1i}, C_{2j} $(1 \le i \le k, 1 \le j \le l+1)$ differs from zero) $D_0 = -XB$ for some $X \in \overline{R}$. Going over from equality (6) to the equality $(A - CX)B + C(D + XB) = 0$ brings us to the possibility already considered. Statement (a) of our proposition holds again.

II. $d(A) = 1$.

We represent the elements A and B in the form (9) and their product in the form (10) and consider several subcases.

II.1. $D_0 = B_0 = 0$. In this case the argument is analogous to that used in case I.1.

II.2 $D_0 = 0$, $B_0 \ne 0$. Comparing the terms beginning with v_i but not with $v_i g_j$ $(1 \le i \le k, 1 \le j \le n)$ and with w_i $(1 \le i \le l+1)$ in both parts of the equality (6), we find that $A_{1i} = A_{2j} = 0$ $(1 \le i \le k, 1 \le j \le l+1)$. Analogously, fixing $i \in \{1, \ldots, k\}$; $s \in \{1, \ldots, m\}$; $p, q \in \{1, \ldots, l+1\}$ $(q \ne l+1)$ and comparing the terms beginning with $v_i, t_i, v_{iq}, u_p, u_{ps}, w_{pq}$ we obtain a system of relations

$$C_{1i} \sum_{j=1}^{n} g_j D_{1j} = 0,$$

$$-C_{1i} \sum_{j=0}^{m} f_j D_{2j} + ({}^{t_i}C_0)D + ({}^{t_i}A)B = 0,$$

$$-C_{1i} \sum_{j=0}^{m} h_{qj} D_{2j} + ({}^{v_{iq}}C_0)D + ({}^{v_{iq}}A)B = 0,$$

$$-C_{2p} \sum_{j=1}^{n} e_j D_{1j} - C_{2p} \sum_{j=0}^{m} f_j D_{2j} + ({}^{u_p}C_0)D + ({}^{u_p}A)B = 0,$$

$$-C_{2p}\sum_{j=1}^{n}e_{sj}D_{1k} - C_{2p}\sum_{j=0}^{m}f_{sj}D_{2j} + (^{u_{ps}}C_0)D + (^{u_{ps}}A)B = 0,$$

$$-C_{2p}\sum_{j=1}^{n}g_{qj}D_{1j} - C_{2p}\sum_{j=0}^{m}h_{qj}D_{2j} + (^{w_{pq}}C_0)D + (^{w_{pq}}A)B = 0.$$

Using the condition $l+m \geq 2$ it is easy to deduce from this that $C_{1i} = 0$, $C_{2p} = 0$ for $p \neq l+1$ (for otherwise, $D = 0$) and $D_{1j} = 0$ ($1 \leq j \leq n$), $D_{2j} = 0$ ($1 \leq j \leq m$). Thus, equality (6) takes the following form:

$$A_0 \cdot B = -(C_{2l+1}w + C_0) \cdot hD_{20} = (C_{2l+1}uf - C_0 h)D_{20}.$$

Comparing the terms that do not begin with u we obtain either $B = \alpha D$, which is impossible (because of the case we are considering), or $A = uA_0'$, $C_0 = uC_0'$, and hence, case (c) holds.

II.3. $D_0 \neq 0$, $B_0 = 0$. In this case the argument is analogous to that used in case II.2.

II.4. $D_0 \neq 0$, $B_0 \neq 0$. Comparing the terms beginning with v_i but not with $v_i g_j$ ($1 \leq i \leq k$, $1 \leq j \leq n$) and with w_i ($1 \leq i \leq l+1$) in both parts of the equality (6) we obtain $C_{1i}D_0 = -A_{1i}B_0$, $C_{2i}D_0 = -A_{2i}B_0$. If at least one of the numbers C_{2i}, A_{2i} ($1 \leq i \leq l+1$) and C_{1i}, A_{1i} ($1 \leq i \leq k$) differs from zero, then $D_0 = -\alpha B_0$, $\alpha \in \mathscr{F}$ and the proof is completed by a transfer from the equality (6) to the equality $(A - \alpha C)B + C(D + \alpha B) = 0$. The remaining possibility leads to the case when $A = C\alpha$, $D = -\alpha B$.

The case when $d(C) = 1$ has been considered completely.

Now assume that $d(C) > 1$. Write the elements A and C in the form of a right linear combination of basic words of length $d(C) - 1$. Let the coefficient at the word σ in the decomposition of C be nonzero. Then $A = \sigma(^\sigma A) + A_0$ and $C = \sigma(^\sigma C) + C_0$, where A_0 and C_0 do not begin with σ, and $d(^\sigma A) = d(^\sigma C) = 1$. From the equality (6) we deduce a system

$$\begin{cases} ^\sigma A \cdot B + ^\sigma C \cdot D = 0, \\ A_0 B + C_0 D = 0 \end{cases}$$

of relations.

Using induction on the number of basic words of length $d(C) - 1$ where at least one element of A or C begins with such a word, we can assume that the equality $A_0 B + C_0 D = 0$ is of the form (7) or (8).

If $A_0 = C_0 Y$, $D = -YB$, then $^\sigma A = ^\sigma CY$, and hence, $A = CY$.

Now if $A_0 = Y(w + u\beta_{21})\lambda$, $B = \mu(h - \beta_{12}f)Z$, $C_0 = Y(u + w\beta_{12})\xi$, $D = \eta(f - \beta_{21}h)Z$, where $\beta_{12}\beta_{21} \neq 1$, then $^\sigma A \cdot \mu(h - \beta_{12}f) + ^\sigma C \cdot \eta(f - \beta_{21}h) = 0$. Since the elements $^\sigma A$ and $^\sigma C$ are of degree 1, this equality is of the form (7) or (8) as well. It is not difficult to verify that in either case the original equality $AB + CD = 0$ is of the form (8). Proposition 2.1 is proved.

Thus, Theorem 2.1 is proved for homogeneous elements A, B, C, D.

Now consider an arbitrary dependence relation (6) and expand the elements A, B, C, and D as sums of homogeneous components:

$$A = A_p + A_{p+1} + \cdots, \quad C = C_q + C_{q+1} + \cdots, \quad (11)$$
$$B = B_r + B_{r+1} + \cdots, \quad D = D_S + D_{S+1} + \cdots.$$

Without loss of generality we may assume that $0 \leq q \leq p$. First, we consider the case $q = 1$ (if $q = 0$, then possibility (1) is realized immediately.)

A homogeneous dependence relation $(A_p | C_1) \cdot \left(\frac{B_r}{D_s}\right) = 0$ should be obtainable from the equalities of the form (1) and (2) mentioned in Theorem 1.1 under the action of a suitable quadruple $(\alpha, \beta, \gamma, \delta)$ of matrices.

If the former of these possibilities holds, it follows from our consideration of the homogeneous case that $A_p = C_1 X_{p-1}$ and $D_S = -X_{p-1} B_r$ for some homogeneous of degree $p - 1$ element X_{p-1}. Therefore, the relation of the form (6) obtained from the original one by action of the quadruple $(1, \beta_0, \gamma_0, 1)$ of matrices:

$$\left(1, \left(\begin{array}{c|c} 1 & 0 \\ \hline -X_{p-1} & 1 \end{array}\right), \left(\begin{array}{c|c} 1 & 0 \\ \hline X_{p-1} & 1 \end{array}\right), 1\right)$$

satisfies the conditions $A \equiv 0 \pmod{\overline{R}^{(p)}}$, $D \equiv 0 \pmod{\overline{R}^{(s)}}$, where $\overline{R}^{(i)} = \{A \in \overline{R} | d(A) > i\}$. Suppose we have found matrices $\beta_i = \left(\begin{array}{c|c} 1 & 0 \\ \hline -X_{p+i+1} & 1 \end{array}\right)$, $\gamma_i = \beta_i^{-1}$, where $d(X_{p+i-1}) \geq p + i - 1$, such that the relation of the form (6) that is obtained from the original one by the action of the quadruple $(1, \beta_0 \cdots \beta_i, \gamma_i \cdots \gamma_0, 1)$ satisfies the conditions

$$A \equiv 0 \pmod{\overline{R}^{(p+i)}}, \quad D \equiv 0 \pmod{\overline{R}^{(s+i)}}.$$

The argument we used to construct the matrices β_0 and γ_0 permits us to assume that matrices β_i and γ_i with the property indicated are found for all $i \in \mathbb{N}$. Setting $\beta = \prod_{i=0}^{\infty} \beta_i$, $\gamma = \beta^{-1}$, we obtain matrices invertible over the ring $\overline{R}$ and such that the action of the quadruple $(1, \beta, \gamma, 1)$ on the original relation transforms it to the form $(0|C) \cdot \left(\frac{B}{0}\right) = 0$. Thus, case (1) of our theorem holds.

Now suppose that $p = q = r = s = 1$, and that the equality $A_1 \cdot B_1 + C_1 \cdot D_1 = 0$ has the form (b) from Proposition 2.1. The above consideration permits us to assume that, under the action of a quadruple $(\alpha_0, \beta_0, \gamma_0, \delta_0)$ of matrices over the field F, the initial dependence relation (6) is transformed into a relation of the same form that satisfies the conditions $A \equiv w \pmod{\overline{R}^{(1)}}$ $B \equiv h \pmod{\overline{R}^{(1)}}$, $C \equiv u \pmod{\overline{R}^{(1)}}$, $D \equiv f \pmod{\overline{R}^{(1)}}$.

Suppose that matrices $\alpha_i, \beta_i, \gamma_i, \delta_i$ from the set of invertible matrices over the ring $\mathscr{F} + \overline{R}^{(i-1)}$ are found such that the relation of the form (6)

that is obtained from the initial one by the action of the quadruple
$$(\alpha_i \cdots \alpha_0, \beta_0 \cdots \beta_i, \gamma_i \cdots \gamma_0, \delta_0 \cdots \delta_i)$$
satisfies the conditions
$$\begin{align} A &\equiv w, & B &\equiv h, \\ C &\equiv u, & D &\equiv f \quad (\mod \overline{R}^{(i+1)}) \end{align} \tag{12}$$
and find the matrices $\alpha_{i+1}, \beta_{i+1}, \gamma_{i+1}, \delta_{i+1}$.

LEMMA 2.1. *Every element $A \in \overline{R}$ can be represented in a unique way in the following forms*:
$$A = wA' + uA'' + A^0 = \overline{A}'h + \overline{A}''f + \overline{A}^0, \tag{13}$$
where the elements $A', A'', A^0, \overline{A}', \overline{A}'', \overline{A}^0$ are reduced, A'' does not begin with f, A^0 with either $u_{l+11}f_{1j}$ $(1 \leq j \leq m)$ or $u_{l+11}e_{ij}$ $(1 \leq j \leq n)$, $\overline{A}''$ does not end with u and $\overline{A}^0$ with either $v_{il}h_{0l}$ $(1 \leq i \leq k)$ or $w_{il}h_{0l}$ $(1 \leq i \leq l)$. Here the products wA' and $\overline{A}'h$ need not be reduced.

PROOF. There is no doubt about the possibility of representing an element in the form (13). We check the uniqueness of the first representation, for example. To this end it suffices to understand that $A' = A'' = A^0 = 0$ if $A = 0$. Let $A' = \sum_{j=0}^m h_j B_j' + \sum_{j=1}^n g_j B_j'' + B_0$ be a reduced representation of the element A'. Then
$$\sum_{j=0}^m \left(uf_j + \sum_{s=1}^m u_{l+1s}f_{sj} \right) B_j' + \sum_{j=1}^n \left(ue_j + \sum_{s=1}^m u_{l+1s}e_{sj} \right) B_j'' + wB_0 + uA'' + A^0 = 0.$$
Comparing those elements of this equality which begin with w, $u_{l+11}f_{1j}$, $u_{l+11}e_{1j}$, and u, we verify successively that $B_0 = 0$, $B_j' = 0$ $(1 \leq j \leq m)$, $B_j'' = 0$ $(1 \leq j \leq n)$, $B_0' = A'' = 0$, $A^0 = 0$. Q.E.D.

We return to the proof of Theorem 2.1. Assuming that the elements A, B, C, and D in (6) are written in the form (11) and satisfy conditions (12), we compute the $(i+3)$rd homogeneous component of the sum $AB+CD$ and equate it to zero;
$$wB_{i+2} + A_{i+2}h + uD_{i+2} + C_{i+2}f = 0.$$
Writing the elements A_{i+2} and C_{i+2} in the form (13) we obtain
$$\begin{cases} B_{i+2} + A'_{i+2}h - h(^f D_{i+2}) + C'_{i+2}f = 0, \\ D_{i+2} - f(^f D_{i+2}) + A''_{i+2}h + C''_{i+2}f = 0, \\ A^0_{i+2}h + C^0_{i+2}f = 0. \end{cases}$$
The last equality of this system means that $A^0_{i+2} = X_{i+1}w$, $C^0_{i+2} = X_{i+1}u$ for some homogeneous of degree $i+1$ element X_{i+1}.

The following chain of equalities and transformations modulo $\overline{R}^{(i+2)}$ determines the desired quadruple $(\alpha_{i+1}, \beta_{i+1}, \gamma_{i+1}, \delta_{i+1})$:

$$(A|C) \cdot \begin{pmatrix} B \\ D \end{pmatrix} \equiv (w + A_{i+2}|u + C_{i+2}) \cdot \begin{pmatrix} h + B_{i+2} \\ f + D_{i+2} \end{pmatrix}$$

$$\equiv ((1 + X_{i+1})w + wA'_{i+2} + uA''_{i+2}|(1 + X_{i+1})u + uC''_{i+2} + wC'_{i+2})$$

$$\cdot \begin{pmatrix} h(1 + {}^fD_{i+2}) - A'_{i+2}h - C'_{i+2}f \\ f(1 + {}^fD_{i+2}) - C''_{i+2}f - A''_{i+2}h \end{pmatrix}$$

$$\downarrow$$

$$((1 + X_{i+1})^{-1}, 1, 1, (1 + {}^fD_{i+2})^{-1})$$

$$\downarrow$$

$$(w(1 + A'_{i+2}) + uA''_{i+2}|u(1 + C''_{i+2}) + wC'_{i+2}) \cdot \begin{pmatrix} (1 - A'_{i+2})h - C'_{i+2}f \\ (1 - C''_{i+2})f - A''_{i+2}h \end{pmatrix}$$

$$\downarrow$$

$$\left(1, \left(\begin{array}{c|c} (1 + A'_{i+2})^{-1} & 0 \\ \hline 0 & (1 + C''_{i+2})^{-1} \end{array}\right), \left(\begin{array}{c|c} 1 + A'_{i+2} & 0 \\ \hline 0 & 1 + C''_{i+2} \end{array}\right), 1\right)$$

$$\downarrow$$

$$(w + uA''_{i+2}|u + wC'_{i+2}) \cdot \begin{pmatrix} h - C'_{i+2}f \\ f - A''_{i+2}h \end{pmatrix}$$

$$\downarrow$$

$$\left(1, \left(\begin{array}{c|c} 1 & -C'_{i+2} \\ \hline -A''_{i+2} & 1 \end{array}\right), \left(\begin{array}{c|c} 1 & -C'_{i+2} \\ \hline -A''_{i+2} & 1 \end{array}\right), 1\right)$$

$$\downarrow$$

$$(w|u) \cdot \begin{pmatrix} h \\ f \end{pmatrix}.$$

Setting $\beta = \prod_{i=0}^{\infty} \beta_i$, $\delta = \prod_{i=0}^{\infty} \delta_i$, $\alpha^{-1} = \prod_{i=0}^{\infty} \alpha_i^{-1}$, $\gamma = \beta^{-1}$ we obtain a quadruple of matrices that reduces the initial relation (6) to the form (2) mentioned in Theorem 2.1.

Thus, we can assume that $\max\{p, q, r, s\} > 1$ in representation (11). Let σ and τ be arbitrary basic words of degrees $q - 1$ and $r - 1$, respectively, and $^\sigma C_q \neq 0$ and $B_r^\tau \neq 0$. Clearly, the relations

$$(^\sigma A|^\sigma C) \cdot \begin{pmatrix} B \\ D \end{pmatrix} = 0, \qquad (^\sigma A|^\sigma C) \cdot \begin{pmatrix} B^\tau \\ D^\tau \end{pmatrix} = 0 \qquad (14)$$

hold.

If the first of these equalities (which satisfies the condition $d(^\sigma C) = 1$) is obtained from a relation of the form (1) mentioned in Theorem 2.1 by means of a suitable quadruple of matrices, then we can assume from the beginning that $B = 0$ and everything is proved.

Otherwise, $d(^\sigma A) = d(^\sigma C) = d(B^\tau) = d(D^\tau) = 1$. Therefore, considering the second of the equalities (14), we can assume that $(^\sigma A|^\sigma C) = (\alpha w | \alpha u)$ for some element α invertible in $\overline{R}$. Then the first of the relations (14) produces the condition $wB + uD = 0$. Representing B and D in the form (11) and using the claim of Theorem 2.1 in the homogeneous case, we obtain $B = hY$ and $D = fY$ for some $Y \in \overline{R}$. Therefore, the initial relation is obtained from $(A|C)(\frac{h}{f}) = 0$ by means of the quadruple $(1, 1, 1, Y)$. It remains to apply the homogeneous case once more, Theorem 2.1 is proved completely.

3. In this subsection we complete the investigation of the properties of the semigroup $\overline{R}^*$ necessary for the proof of its embeddability in a group.

Let Γ denote the group of invertible elements of the ring $\overline{R}$, and let $\mathfrak{A}$ be the subsemigroup of $\overline{R}^*$ generated by all components of the vectors $(w|u)\beta$ and $\beta(\frac{h}{f})$, where $\beta \in \mathrm{GL}_2(\overline{R})$, and by the set Γ. Set $\mathscr{L} = \overline{R}^* \setminus (\overline{R}^*(\mathfrak{A} \setminus \Gamma)\overline{R}^*)$.

THEOREM 2.2. *The following statements hold*:
(1) *The semigroup $\mathfrak{A}$ is generated by the set $\Gamma \cup P$, where*

$$P = \{w + uA, u + wA, f + Ah, h + Af | A \in \overline{R}\}.$$

(2) *The set $\mathfrak{B}$ is a subsemigroup of $\overline{R}^*$ and each dependence relation of the form (6) in which an element $b \in \mathfrak{B}$ occurs satisfies either the condition $A = CX$ or $C = AX$ for some $X \in \overline{R}^*$.*

(3) *The semigroup $\overline{R}^*$ is a free product of semigroups $\mathfrak{A}$ and $\mathfrak{B}$ with the subgroup Γ amalgamated.*

PROOF. (1) Let A be the first component of the vector $(w|u)\beta$, where

$$\beta = \begin{pmatrix} \beta_{11} & \beta_{12} \\ \beta_{21} & \beta_{22} \end{pmatrix} \in \mathrm{GL}_2(\overline{R}).$$

At least one of the elements β_{11} and β_{21} is invertible in $\overline{R}$. Therefore, either $A = (w + u\beta_{21}\beta_{11}^{-1})\beta_{11}$ or $A = (u + w\beta_{11}\beta_{21}^{-1})\beta_{21}$. For other generators of the semigroup $\mathfrak{A}$ the argument is analogous.

(2) Suppose that $A_1, A_2 \notin \overline{R^*}P\overline{R^*}$ but $A_1 \cdot A_2 = U_p V$, where $U, V \in \overline{R^*}$, $p \in P$. Clearly, the dependence relation $(A_1| - U_p) \cdot \binom{A_2}{V} = 0$ cannot satisfy case (2) of Theorem 2.1. Therefore, $(A_1| - U_p) = (\alpha\beta_{11}|\alpha\beta_{12})$ for certain $\alpha \in \overline{R^*}$ and an invertible matrix $\beta \in \mathrm{GL}_2(\overline{R})$. Since at least one of the elements β_{11} and β_{12} is invertible, we can assume without loss of generality that $A_1 = U_p X$ or $U_p = A_1 X$ for some $X \in \overline{R^*}$. Due to the choice of the element A_1, only the latter of these possibilities can hold. Then $A_2 = XV$, but analogous considerations applied to the dependence relation $(A_1| - U) \cdot \binom{X}{p} = 0$ show that $X \in \overline{R^*}P$ contrary to the choice of the element A_2. Thus, it is proved that $\mathfrak{B}$ is a subsemigroup of $\overline{R^*}$. The second part of the statement (2) is obvious.

(3) First of all, we verify that the semigroups $\mathfrak{A}$ and $\mathfrak{B}$ generate $\overline{R^*}$. It is clear that all elements of degree zero from $\overline{R^*}$ belong to $\Gamma = \mathfrak{a} \cap \mathfrak{B}$. Now if $z \in \overline{R^*} \setminus (\mathfrak{A} \cup \mathfrak{B})$, then $z = X_p \gamma$ or $z = \gamma_p X$, where $\gamma \in \Gamma$, $p \in P$, $X \in \overline{R}$ and $d(X) < d(z)$. It remains to apply induction on the degree.

Now fix in $\mathfrak{B}$ a set Q of atoms of the semigroup $\overline{R^*}$ such that $Q \cup \Gamma$ generates $\mathfrak{B}$ and prove that every relation in $\overline{R^*}$ of the form $\gamma_0, \xi_1, \gamma_1 \cdots \gamma_{n-1}\xi_n\gamma_n = \delta_0\eta_1\delta_1\cdots\delta_{m-1}\eta_m\delta_m$ with $n \leq m$, $\gamma_i, \delta_j \in \Gamma$ ($0 \leq i \leq n$, $0 \leq j \leq m$); $\xi_i, \eta_j \in P \cup Q$ ($1 \leq i \leq n$, $1 \leq j \leq m$) is a consequence of the defining relations in the semigroups $\mathfrak{A}$ and $\mathfrak{B}$, while $n = m$. We carry out induction on m. Since $\mathfrak{A} \cap \mathfrak{B} = \Gamma$, we can assume that $n > 1$. Let

$$A = \gamma_0\xi_1\cdots\xi_{n-1}\gamma_{n-1}, \qquad B = \xi_n\gamma_n,$$
$$C = \delta_0\eta_1\cdots\eta_{m-1}\delta_{m-1}, \qquad D = -\eta_m\delta_m.$$

Then the dependence relation $(A|C) \cdot \binom{B}{D} = 0$ holds, and Theorem 2.1 makes it possible to assume that either $C = AX$ and $B = XD$ or $A = Xp_1\alpha_1$, $B = \alpha_2 p_2 Y$, $C = Xp_3\alpha_3$, $D = \alpha_4 p_4 Y$, and hence, the induction hypothesis on the relation $C = AX$ produces the result desired. In the latter case apply the induction hypothesis first to the equality $A = Xp_1\alpha_1$ and then to the equality $C = Xp_3\alpha_3$. Theorem 2.2 is proved.

§2. Embedding the semigroup $\overline{R^*}$ in a group

1. Recall that a semigroup S is called *rigid* if, for every equality of the form $ab = cd$ ($a, b, c, d \in S$), either $a = cx$ and $d = xb$ or $c = ax$ and $b = xd$ for some $x \in S$. We say that S is a *cancellative semigroup* if every equality of the form $xy = xz$ or $yx = zx$ ($x, y, z \in S$) implies the condition $y = z$.

The following theorem proved by R. Doss [36] gives necessary conditions for embeddability in a group.

THEOREM (R. Doss). *A rigid cancellative semigroup is embeddable in a group.*

Statement (2) of Theorem 2.2 shows that $\mathfrak{B}$ is a rigid cancellative semigroup. By the Doss theorem, it is embeddable in a universal group $G(\mathfrak{B})$. Assume that the semigroup $\mathfrak{A}$ is embeddable in a universal group $G(\mathfrak{A})$ as well. Then as it follows from Theorem 2.2, the group $G(\mathfrak{A}) *_\Gamma G(\mathfrak{B})$, the free product of the groups $G(\mathfrak{A})$ and $G(\mathfrak{B})$ with the subgroup Γ amalgamated, contains $\overline{R^*}$. Thus, it suffices to prove that the semigroup $\mathfrak{A}$ is embeddable in a group.

The main goal of this section is to prove the following theorem.

THEOREM 2.3. *The semigroup $\mathfrak{A}$ is embeddable in a rigid cancellative semigroup.*

The ambient rigid semigroup will be constructed from a certain ring R_τ which is considered in part 2 of this section.

2. Consider an algebra R_τ with identity and defining relations of the form

$$\left(\begin{array}{c|cc|cc} V & 'v & & 't & T \\ \hline W & 'w & & 'u & U \\ 0 & -u\tau & & u & i' \end{array}\right) \cdot \left(\begin{array}{cc|c} 'h & H & G \\ h & h' & g' \\ \hline \tau h & f' & e' \\ 0 & F & E \end{array}\right) = \left(\begin{array}{cc|c} 0 & 0 & * \\ 0 & 0 & 0 \\ 0 & 0 & 0 \end{array}\right). \tag{16}$$

Order the letters occurring in the equalities (16) in such a way that the elements t_i ($1 \leq i \leq k$) and u_j ($1 \leq j \leq l$) are the greatest elements in their rows. Then again,

$$u\tau h_i \ (1 \leq i \leq m), \quad u\tau g_i \ (1 \leq i \leq n), \quad u_i\tau h \ (1 \leq i \leq l),$$
$$u_i f_j \ (1 \leq i \leq l, 1 \leq j \leq m), \quad u_i e_j \ (1 \leq i \leq l, \ 1 \leq j \leq n),$$
$$t_i \tau h \ (1 \leq i \leq k), \quad t_i f_j \ (1 \leq i \leq k, \ 1 \leq j \leq m)$$

will be the senior elements. Exactly as for the ring R, we introduce the notions of basic words and reduced elements.

In the usual way we define on R_τ the lower degree function $d_\tau \colon R_\tau \to \mathbb{N} \cup \{+\infty\}$ which takes value one on each letter from (16) and satisfies the conditions:

(1) $d_\tau(A) = +\infty \Leftrightarrow A = 0$,

(2) $d_\tau(A + B) \geq \min\{d_\tau(A), d_\tau(B)\}$ for any elements $A, B \in R_\tau$.

In contradistinction to the case of the ring R, the function we have introduced does not satisfy the condition $d_\tau(A \cdot B) \geq d_\tau(A) + d_\tau(B)$, which fails, for example, for $A = u$ and $B = \tau h_1$. Nevertheless, we have

LEMMA 2.2. *For any elements $A, B \in R_\tau$ the following inequality holds*:

$$d_\tau(A \cdot B) \geq d_\tau(A) + d_\tau(B) - 1. \tag{17}$$

PROOF. Clearly, it suffices to verify the inequality (17) for homogeneous elements A and B which, in this case, can be written in the following form:

$$A = \sum_{i+1}^{l+1} A_{1i}u_i\tau + \sum_{i=1}^{k} A_{2i}t_i\tau + \sum_{i=1}^{l+1} A_{3i}u_i + \sum_{i=1}^{k} A_{4i}t_i + A_5,$$

$$B = \sum_{j=0}^{m} \tau h_j B_{1j} + \sum_{j=1}^{n} \tau g_j B_{2j} + \sum_{j=0}^{m} h_j B_{3j} + \sum_{j=1}^{n} g_j B_{4j} + B_5, \quad (18)$$

where A_5 does not end with $u_i\tau$, u_i $(1 \leq i \leq l+1)$, $t_i\tau$, t_i $(1 \leq i \leq k)$ and B_5 does not begin with τh_j, h_j $(0 \leq j \leq m)$, τg_j, g_j $(1 \leq j \leq n)$. Then

$$A \cdot B = \left(\sum_{i=1}^{l} A_{1i}u_i\tau + \sum_{i=1}^{k} A_{2i}t_i\tau + \sum_{i=1}^{l} A_{3i}u_i + \sum_{i=1}^{k} A_{4i}t_i + A_5 \right)$$

$$\cdot \left(\sum_{j=1}^{m} \tau h_j B_{1j} + \sum_{j=1}^{n} \tau g_j B_{2j} + \sum_{j=1}^{m} h_j B_{3j} + \sum_{j=1}^{n} g_j B_{4j} + B_5 \right)$$

$$+ A_{1l+1}u\tau \left(\sum_{j=0}^{m} \tau h_j B_{1j} + \sum_{j=1}^{n} \tau g_j B_{2j} + hB_{30} + B_5 \right)$$

$$+ A_{3l+1}u \left(\tau h B_{10} + \sum_{j=0}^{m} h_j B_{3j} + \sum_{j=1}^{n} g_j B_{4j} + B_5 \right)$$

$$+ \left(\sum_{i=1}^{l} A_{3i}u_i + \sum_{i=1}^{k} A_{4i}t_i + A_5 \right) hB_{30}$$

$$+ \left(\sum_{i=1}^{l} A_{1i}u_i\tau + \sum_{i=1}^{k} A_{21}t_i\tau + A_5 \right) \tau h B_{10}$$

$$+ A_{1l+1} \left[\sum_{j=1}^{m} \left(uf_j + \sum_{s=1}^{j} u_{l+1s}f_{sj} \right) B_{3j} + \sum_{j=1}^{n} \left(ue_j + \sum_{s=1}^{m} u_{l+1s}e_{sj} \right) B_{4j} \right]$$

$$+ A_{3l+1} \left[\sum_{j=1}^{m} \left(uf_j + \sum_{s=1}^{j} u_{l+1s}f_{sj} \right) B_{ij} + \sum_{j=1}^{n} \left(ue_j + \sum_{s=1}^{m} u_{l+1s}e_{sj} \right) B_{2j} \right]$$

$$- \left[\sum_{i=1}^{l} A_{3i} \left(w_i h + \sum_{s=i}^{l} w_{is}h_{s0} \right) + \sum_{i=1}^{k} A_{4i} \left(v_i h + \sum_{s=1}^{l} v_{is}h_{s0} \right) \right] B_{10}$$

$$- \left[\sum_{i=1}^{l} A_{1i} \left(w_i h + \sum_{s=i}^{l} w_{is}h_{s0} \right) + \sum_{i=1}^{k} A_{2i} \left(v_i h + \sum_{s=1}^{l} v_{is}h_{s0} \right) \right] B_{30}.$$
(19)

Each of the first five summands in this sum either equals zero or is of degree $d\tau(A) + d\tau(B)$. The remaining summands are reduced, and if they do not equal zero, they are of degree $d_\tau(A) + d_\tau(B) - 1$. Thus, the inequality $d_\tau(A \cdot B) \geq d_\tau(A) + d_\tau(B) - 1$ is proved.

Lemma 2.2 is proved.

The inequality (17) makes it possible to introduce on R_τ a norm by the formula $\|A\| = e^{-d_\tau(A)+1}$. Let $\overline{R}_\tau$ denote the completion of the ring R_τ with respect to this norm. It can be identified with the set of formal power series $\sum_{i=0}^\infty A^{(i)}$, where the $A^{(i)}$ are d_τ-homogeneous elements of degree i of the ring R_τ with componentwise addition and multiplication of the homogeneous components, subordinated to formula (19).

The inequality (17) shows, in particular, that those and only those elements A are invertible in the ring $\overline{R}_\tau$ for which $d_\tau(A) = 0$. Indeed, it is clear that the condition $d_\tau(A) = 0$ is necessary for the invertibility. The converse follows from the fact that $d_\tau(A^n) \geq n(d_\tau(A) - 1)$. Thus, for every $A \in \overline{R}_\tau$ that satisfies the condition $d_\tau(A) > 1$, the element $1 + \sum_{i=1}^\infty A^i \in \overline{R}_\tau$ is the inverse for the element $1 - A$. Now if $d_\tau(A) = 1$, then it suffices to consider the element A^2 and use the formula $1 - A^2 = (1+A)(1-A)$.

Let the symbol Γ_τ denote the group of invertible elements of the ring $\overline{R}_\tau$ and set $P_\tau = \{u, h, \tau + A | A \notin \Gamma_\tau, {}^\tau A = A^\tau = 0\}$.

LEMMA 2.3. *The set P_τ consists of nondivisors of zero.*

PROOF. First of all observe that, for the series $A = \sum_{i=p}^\infty A^{(i)}$ and $B = \sum_{j=q}^\infty B^{(j)} \in \overline{R}_\tau$ that satisfy the condition $A \cdot B = 0$, the following relations must hold by Lemma 2.2:

$$d_\tau(A^{(p)} \cdot B^{(q)}) = p + q, \quad d_\tau(A^{(p)} \cdot B^{(q+1)} + A^{(p+1)} \cdot B^{(q)} + A^{(p)} \cdot B^{(q)}) > p + q. \tag{20}$$

Thus, to prove Lemma 2.3, it suffices to verify that for the elements of the set P_τ the corresponding systems of the form (20) are inconsistent.

First assume that $u \cdot B = 0$ and $B^{(q)} \neq 0$. Writing the elements $B^{(q)}$ and $B^{(q+1)}$ in the form (18) and using the formula (19) we obtain that the system (20) is equivalent to the following relations:

$$\sum_{j=1}^m \left(uf_j + \sum_{s=1}^j u_{l+1s}f_{sj} \right) B_{1j}^{(q)} + \sum_{j=1}^n \left(ue_j + \sum_{s=1}^m u_{l+1s}e_{sj} \right) B_{2j}^{(q)} = 0,$$

$$u(\tau h B_{10}^{(q)} + \sum_{j=0}^m h_j B_{3j}^{(q)} + \sum_{j=1}^n g_j B_{4j}^{(q)} + B_5^{(q)} + \sum_{j=1}^m \left(uf_j + \sum_{s=1}^j B_{1j}^{(q+1)} u_{l+1s}f_{sj} \right)$$

$$+ \sum_{j=1}^n \left(ue_j + \sum_{s=1}^m u_{l+1s}e_{sj} \right) B_{2j}^{(q+1)} = 0.$$

The first of these equalities shows that $B_{1j}^{(q)} = 0$ $(1 \leq j \leq m)$ and $B_{2j}^{(q)} = 0$ $(1 \leq j \leq n)$, and it follows from the second that $B_{10}^{(q)} = 0$, $B_{3j}^{(q)} = 0$

$(0 \leq j \leq m)$, $B_{4j}^{(q)} = 0$ $(1 \leq j \leq n)$, $B_{1j}^{(q+1)} = 0$ $(1 \leq j \leq m)$, $B_{2j}^{(q+1)} = 0$ $(1 \leq j \leq n)$ and $B_5^{(q)} = 0$, contrary to our assumption.

It is obvious that u is not a right zero divisor because the product $B \cdot u$ is reduced. Thus, u is not a zero divisor in $\overline{R}_\tau$.

Switching to the ring that is anti-isomorphic to $\overline{R}_\tau$, we obtain that h is not a zero divisor in $\overline{R}_\tau$.

If $(\tau + A) \cdot B = 0$, then comparing in this equality the terms beginning with τ, we obtain $B = 0$. Analogously, we verify that $\tau + A$ is not a right zero divisor.

Lemma 2.3 is proved.

3. In this part we construct an embedding $\varphi \colon \overline{R} \to \overline{R}_\tau$.

First of all, we reorder the letters occurring in the defining relations of the ring R so that the words

$$wh_i \ (0 \leq i \leq m), \quad wg_i \ (1 \leq i \leq n), \quad u_i f_j \ (1 \leq i \leq l, \ 0 \leq j \leq m),$$
$$u_i e_j \ (1 \leq i \leq l, \ 1 \leq j \leq n), \quad t_i f_j \ (1 \leq i \leq k, \ 0 \leq j \leq m)$$

become senior.

Next we consider a mapping φ that transforms the entries of the matrices from (3) into corresponding entries of the matrices occurring in the relations (16). It is clear that φ can be extended up to a homomorphisms of the algebras $R \to R_\tau$, with the basic words being mapped into basic words under the action of φ.

LEMMA 2.4. *The homomorphism $\varphi \colon R \to R_\tau$ is an embedding.*

PROOF. Our argument follows Koshevoĭ [17]. Let A be an element of the least upper degree that belongs to $\ker \varphi$. It is clear that we can restrict ourselves to the case when A has the following representation in the form (13): $A = uA_1 + wA_2$. We obtain $0 = A^\varphi = uA_1^\varphi + u\tau A_2^\varphi$. By Lemma 2.3, $A_1^\varphi + \tau A_2^\varphi = 0$. Writing the element A_1 in the form $A_1 = fA_1' + A_1''$, where A_1'' does not begin with f, we see that $(A_1'')^\varphi = 0$ and $h(A_1')^\varphi + A_2^\varphi = 0$. Taking into consideration that A_1 does not begin with f and that A has the minimal upper degree among the elements of the kernel of the mapping φ, we obtain $A_1' = A_1'' = A_2 = 0$. Q.E.D.

Define on R a new function of lower degree $d' \colon R \to \mathbb{N} \cup \{+\infty\}$ that equals 2 on w and f and 1 on the remaining letters from (3). It is clear that every element $A \in \overline{R}$ can be uniquely represented as a formal sum of d'-homogeneous components that are mapped by φ into d_τ-homogeneous components of the same degree. This makes it possible to extend φ to an embedding of algebras $\varphi \colon \overline{R} \to \overline{R}_\tau$.

Let $\mathfrak{A}_\tau$ denote the subsemigroup of $\overline{R}_\tau$ that is generated by the set $\Gamma_\tau \cup P_\tau$. Obviously, $\mathfrak{A}^\varphi \subseteq \mathfrak{A}_\tau$. It remains to verify that $\mathfrak{A}_\tau$ is a rigid cancellative semigroup. Here $\mathfrak{A}_\tau$ consists of nondivisors of zero, by Lemma 2.3.

LEMMA 2.5. $\mathfrak{A}_\tau$ *is a rigid cancellative semigroup.*

PROOF. Each element of $\mathfrak{A}_\tau$ can be presented in the form

$$\gamma_0 p_1 \gamma_1 \cdots \gamma_{n-1} p_n \gamma_n,$$

where $n \geq 0$, $\gamma_i \in \Gamma_\tau$, $p_i \in P_\tau$.

Clearly, for the proof of the lemma it suffices, by Lemma 2.3, to show that each equality of the form

$$pA = \gamma q B, \tag{21}$$

where $\gamma \in \Gamma_\tau$, $p, q \in P_\tau$, $A, B \in \mathfrak{A}_\tau$, implies the relation $\gamma_q = p\delta$ for some $\delta \in \Gamma_\tau$.

We consider consecutively several cases.

I. $p = u$. Present the element γ in the form $\gamma = \alpha + D$, where $\alpha \in F^*$, $d_\tau(D) \geq 1$.

If $q = h$ ($q = \tau + P \in \mathscr{P}_\tau$), then comparing the terms of (21) beginning with $h(\tau)$, we obtain the equality $(\alpha + ({}^h D)h)B = 0$ $((\alpha + ({}^\tau D)(\tau + P))B = 0)$ which contradicts our choice of the element B.

Thus, $q = u$. Write the elements A and B in a reduced form

$$A = \sum_{i=1}^m \tau h_i A_{1i} + \sum_{i=1}^n \tau g_i A_{2i} + A_0,$$

$$B = \sum_{i=1}^m \tau h_i B_{1i} + \sum_{i=1}^n \tau g_i B_{2i} + B_0,$$

where A_0 and B_0 do not begin with τh_i ($1 \leq i \leq m$), τg_i ($1 \leq i \leq n$). Then the equality (21) takes the form

$$u\left(\sum_{i=1}^m f_i A_{1i} + \sum_{i=1}^n e_i A_{2i} + A_0\right) + \sum_{s=1}^m u_{l+1s}\left(\sum_{i=1}^m f_{si} A_{1i} + \sum_{i=1}^n e_{si} A_{2i}\right)$$

$$= (\alpha + D)\left[u\left(\sum_{i=1}^m f_i B_{1i} + \sum_{i=1}^n e_i B_{2i} + B_0\right)\right.$$

$$\left. + \sum_{s=1}^m u_{l+1s}\left(\sum_{i=1}^m f_{si} B_{1i} + \sum_{i=1}^n e_{si} B_{2i}\right)\right].$$

It can be seen from it that D has a reduced representation $D = u D_0 + \sum_{s=1}^m u_{l+1s} D_s$ and the following condition holds:

$$\sum_{i=1}^m f_{si} A_{1i} + \sum_{i=1}^n e_{si} A_{2i} = \alpha\left(\sum_{i=1}^m f_{si} B_{1i} + \sum_{i=1}^n e_{si} B_{2i}\right) + D_s u B.$$

Therefore, D_S can be represented in the form $D_S = \sum_{i=1}^m f_{si} D_{si} + \sum_{i=1}^n e_{si} \overline{D}_{si}$ and $D_{si} u B = A_{1i} - \alpha B_{1i}$, $\overline{D}_{si} u B = A_{2i} - \alpha B_{2i}$. Analogously we obtain

$$\sum_{i=1}^m f_i A_{1i} + \sum_{i=1}^n e_i A_{2i} + A_0 = \alpha\left(\sum_{i=1}^m f_i B_{1i} + \sum_{i=1}^n e_i B_{2i} + B_0\right) + D_0 u B.$$

Taking into consideration previous relations we conclude that

$$\sum_{i=1}^{m} f_i D_{si}, uB + \sum_{i=1}^{n} e_i \overline{D}_{si} uB - D_0 uB = \alpha B_0 - A_0.$$

Therefore,

$$DuB = uA - \alpha uB$$
$$= \sum_{i=1}^{m} u\tau h_i D_{si} uB + \sum_{i=1}^{n} u\tau g_i \overline{D}_{si} uB$$
$$+ u \left(D_0 - \sum_{i=1}^{m} f_i D_{si} - \sum_{i=1}^{n} e_i \overline{D}_{si} \right) uB.$$

Finally, we obtain $D = uK$ for some $K \in \overline{R}_\tau$ and $\gamma_u = u(\alpha + Ku)$.

II. $p = h$. Analogously to the previous case we confirm that $q = h$. It follows from (21) that $h(A - \alpha B) = DhB$. Therefore, $D = hK$ and $\gamma h = h(\alpha + Kh)$.

III. $p = \tau + P \in \mathscr{P}_\tau$. Our previous considerations show that $q = \tau + Q \in P_\tau$. Comparing the terms of the equality (21) that begin and not begin with τ we obtain the relations

$$A = (\alpha + K(\tau + Q))B, \qquad PA = (\alpha Q + K_1(\tau + Q))B,$$

where $D = \tau K + K_1$ and K_1 does not begin with τ. Substituting into the latter equality the value of A from the former one we obtain

$$\alpha(P - Q) = (K_1 - PK)(\tau + Q).$$

Since P and Q do not end with τ, we obtain $P = Q$, $K_1 = PK$, that is, $D = (\tau + P)K$ and $\gamma(\tau + P) = (\tau + P)(\alpha + K(\tau + P))$. Q.E.D.

Lemma 2.5 and with it Theorem 2.3 are completely proved.

5. Now we prove that the algebra R_0 without identity over the field $\mathscr{F}$ given by the following defining relations

$$\left(\begin{array}{cc|cc} V & 'v & 't & T \\ \hline 1+W & 'w & 'u & U \\ 0 & 1+w & u & u' \end{array} \right) \cdot \left(\begin{array}{cc|c} 'h & H & G \\ h & h' & g' \\ \hline 1+f & f' & e' \\ 0 & 1+F & E \end{array} \right) = \left(\begin{array}{cc|c} 0 & 0 & * \\ 0 & 0 & 0 \\ \hline 0 & 0 & 0 \end{array} \right), \quad (22)$$

where the identity is used only formally, is not quasi-invertible but its adjoined semigroup is embeddable in a group.

Only the second statement needs to be proved. Let $R_0^\#$ denote the algebra obtained from R_0 by adjoining an external identity. Shirshov's lemma ([2, vol. I, p. 53]) shows that, as a base of $R_0^\#$ we can take all words that do not contain subwords of the form

$$v_i h_j \ (1 \leq i \leq k, \ 0 \leq j \leq m), \qquad w_i h_j \ (1 \leq i \leq l+1, \ 0 \leq j \leq m),$$
$$w_i g_j \ (1 \leq i \leq l+1, \ 1 \leq j \leq n).$$

Consider the mapping ψ that transforms every entry of the matrices from (3) into a corresponding entry of the matrices occurring in the relation (22). Clearly, it can be extended up to an epimorphism $\psi: R \to R_0^\#$. Moreover, it is not difficult to prove that ψ is an isomorphism, that is $R \cong R_0^\#$. Furthermore, the mapping $A \to 1+A$ is an embedding of the adjoined semigroup of the ring R_0 into the multiplicative semigroup of the ring $R_0^\#$ that is isomorphic to $R^\#$. Since the semigroup R^* is embeddable in a group, the adjoined semigroup of the ring R_0 is embeddable in the same group. Q.E.D.

In conclusion, we note that the example of Bokut' [4]–[6] remains up to now the only example of a noninvertible ring that is embeddable in a group and that is also a semigroup algebra. Thus, the problem of the construction of a noninvertible semigroup algebra over an arbitrary field for which the embeddability in a group could be proved comparatively simply remains unsolved, in contradistinction to the works [4]–[6].

We would like to draw attention to the following problem as well: does there exist a noninvertible ring without zero divisors and such that for every $n \in \mathbb{N}$ the semigroup P_n of upper-triangular matrices over it of order n with nonzero entries on the main diagonal is embeddable in a group? For the examples given we have proved that P_1 is embeddable in a group while the ring itself is not invertible.

Chapter 3
Representation of Finite-Dimensional Lie Algebras in Radical Rings

1. Throughout this chapter all rings considered are assumed, unless otherwise stipulated, associative without identity and algebras are considered over a fixed field F. All necessary information on Lie algebras and their universal envelopes can be found, e.g., in the book [14].

Let R be a ring and $R^{(-)}$ its adjoined Lie ring (that is, the set R with the operation of addition and the operation of multiplication $[xy] = xy - yx$). We say that a Lie ring $\mathfrak{L}$ is *representable in the ring* R if there exists an embedding $\mathfrak{L} \to R^{(-)}$.

Cohn [28] proved that the universal envelope of an arbitrary Lie algebra is embeddable in a division ring. Thus, every Lie algebra possesses a representation in an associative division ring. He produced an example of a (3-dimensional simple) Lie algebra whose universal envelope is not embeddable in a radical ring. L. A. Bokut' raised the problem of finding conditions necessary and sufficient for the representability of a Lie algebra in radical rings.

The main result of this chapter is

THEOREM 3.1. *The following conditions are equivalent for a finite-dimensional Lie algebra $\mathfrak{L}$:*
  (a) *The universal envelope $U(\mathfrak{L})$ is embeddable in a radical ring.*
  (b) *The Lie algebra $\mathfrak{L}$ possesses a representation in a radical ring.*
  (c) *The algebra $\mathfrak{L}$ is solvable.*

The undecidability of the word problem in the class of radical rings will be deduced from this theorem (part 4).

Note that Theorem 3.1 can be used, as observed by V. N. Gerasimov, to construct simple examples of Ore domains that do not possess a universal division ring of quotients (see [8, p. 1299]).

2. The proof of the theorem will follow the scheme (a)⇒(b)⇒(c)⇒(a). The first implication in this chain is obvious.

An important role in our argument will be played by the following condition (∗) on a Lie ring $\mathfrak{L}$:

(∗) Every nonzero finitely generated subring of the Lie ring $\mathfrak{L}$ differs from its commutator subring.

For a finite-dimensional Lie algebra this condition is obviously equivalent to the solvability. Therefore, the implication (b) ⇒(c) follows from Lemma 3.1.

LEMMA 3.1. *Let R be a radical ring. Then the adjoined Lie ring $R^{(-)}$ satisfies the condition (∗).*

PROOF. Let $x_1, \ldots, x_n$ be generators of a nonzero Lie ring $\mathfrak{L} \subseteq R^{(-)}$. Assume, contrary to what we are proving, that $[\mathfrak{L}, \mathfrak{L}] = \mathfrak{L}$. Then an arbitrarily chosen generator x_k $(1 \leq k \leq n)$ can be represented as a combination with integer coefficients of the ring commutators of $x_1, \ldots, x_n$. Each of these commutators can be written as a (noncommutative) polynomial of $x_1, \ldots, x_n$ and then pull together the monomials beginning with the same letter. As a result we obtain an equality of the form $x_k + \sum_{s=1}^{n} x_s p_{ks}(x_1, \ldots, x_n) = 0$, where the p_{sk} are polynomials with integer coefficients without a constant term. Denoting by P the matrix over R of order n formed by the elements $p_{sk}(x_1, \ldots, x_n)$ and by X the vector row $(x_1, \ldots, x_n)$, we arrive at the relation $X(1 + P) = 0$, which implies $X = 0$ in a radical ring. This contradicts the choice of the ring $\mathfrak{L}$. Lemma 3.1 is proved.

3. Finally, to prove the implication (c)⇒(a) we show that the universal envelope $U(\mathfrak{L})$ of a solvable Lie algebra $\mathfrak{L}$ (without the assumption of finite dimensionality) is embeddable in a radical ring. It suffices to verify that the set of all square matrices over $U(\mathfrak{L})$ is potentially quasi-invertible (see Corollary 2 to Theorem 1.4). In particular, this means that the following condition must hold in the algebra $U(\mathfrak{L})$:

(∗∗) For any two square matrices X and P of the same size if either $X(1 + P) = 0$ or $(1 + P)X = 0$ then $X = 0$.

Besides that, it is known [22] that the universal envelope of a solvable Lie algebra possesses a classical division ring of quotients over which all matrices which are not zero divisors are invertible. Thus, it remains to verify that condition (∗∗) follows from the solvability of the Lie algebra $\mathfrak{L}$. This follows from Proposition 3.1.

PROPOSITION 3.1. *A universal envelope $U(\mathfrak{L})$ of a Lie algebra $\mathfrak{L}$ satisfies condition (∗∗) if and only if condition (∗) holds in the algebra $\mathfrak{L}$.*

PROOF. The necessity follows from the proof of Lemma 3.1. Before proving sufficiency, we introduce certain notations.

Let $\mathfrak{L}$ be a Lie algebra, $\mathscr{H} = \{h_j, j \in J\}$ a basis of the commutant subalgebra $[\mathfrak{L}, \mathfrak{L}]$, and let $\mathscr{F} = \{f_k, k \in K\}$ be its extension to a basis of

$\mathfrak{L}$. We assume that the sets J and K are well ordered. By the Birkhoff-Witt theorem (see, e.g., [2, vol. II, p. 51]), a basis of the algebra $U(\mathfrak{L})$ is formed by all words of the form

$$h_{j_1}^{\alpha_1} \cdots h_{j_s}^{\alpha_s} f_{k_1}^{B_1} \cdots f_{k_t}^{B_t}, \tag{1}$$

where

$$j_1 < \cdots < j_s, \ k_1 < \cdots < k_t, \ \alpha_p, \beta_q \in \mathbb{N} \ (1 \le p \le s, \ 1 \le q \le t).$$

For convenience' sake we use multi-index notation. For example, an expression of the form (1) will be denoted by $h^\alpha f^\beta$, where $\alpha \in \mathbb{N}^J$ and $\beta \in \mathbb{N}^K$. Let $\operatorname{supp}(\beta) = \{k \in K \mid \beta(k) \neq 0\}$ and call this set the *carrier* of β. We consider only multi-indexes with finite carriers. Thus, the number $|\beta| = \sum_{k \in K} \beta(k)$ is always defined. Also, we partially order the multi-indexes considering that $\beta \le \gamma$ if $\operatorname{supp}(\beta) \subset \operatorname{supp}(\gamma)$ and $\beta(k) \le \gamma(k)$ for every $k \in \operatorname{supp}(\beta)$. The number $\deg(h^\alpha f^\beta) = |\alpha| + |\beta|$ is called the *degree* of the word (1). The notion of degree is naturally extended to the set of all matrices over $U(\mathfrak{L})$ and every matrix P over $U(\mathfrak{L})$ can be written in a unique way in the form

$$P = \sum_\beta P_\beta(h) f^\beta, \tag{2}$$

where $P_\beta(h)$ is a (noncommutative) matrix polynomial of the elements of the set $\mathscr{H}$.

The following lemma is probably known and is given for the reader's convenience only.

LEMMA 3.2. *For any multi-indexes α and β and a matrix polynomial $P(h)$ of the elements of the set $\mathscr{H}$ there exist uniquely determined polynomials $S_\gamma^{\alpha,\beta}(h)$ and matrix polynomials $T_\gamma^\alpha(P;h)$ $(\gamma \in \mathbb{N}^K)$ of the elements of the set $\mathscr{H}$ such that*

$$f^\alpha f^\beta = f^{\alpha+\beta} + \sum_{\gamma < \alpha+\beta} S_\gamma^{\alpha\beta}(h) f^\gamma, \qquad f^\alpha P(h) = P(h) f^\alpha + \sum_{\gamma < \alpha} T_\gamma^\alpha(P;h) f^\gamma.$$

Here the polynomials $S_\gamma^{\alpha\beta}(h)$ and $T_\gamma^\alpha(P;h)$ have no constant terms and

$$\deg T_\gamma^\alpha(P,h) \le \deg P(h).$$

PROOF. The proof is given only for the polynomials $\deg T_\gamma^\alpha(P;h)$ since analogous argument can be carried out in the remaining case as well.

First of all, observe that it is sufficient to prove the claim only for $P(h) = h^\beta$ (the general case follows from here by linearity). Therefore, we use double induction on the pair $(|\alpha|, |\beta|)$ of numbers.

If $|\alpha| = |\beta| = 1$, then everything follows from the formula

$$f^\alpha h^\beta = f_k h_l = h_l f_k + [f_k, h_l] = h^\beta f^\alpha + T_0^\alpha(h^\beta; h) f^0,$$

where $\operatorname{supp}(\alpha) = \{k\}$, $\operatorname{supp}(\beta) = \{l\}$, $T_0^\alpha(h^\beta, h) = [f_k, h_l]$.

Assume that the claim we are proving holds for $|\alpha| = 1$, $|\beta| < n$. Let $|\alpha| = 1$, $|\beta| < n$ and $\mathrm{supp}(\alpha) = \{k\}$, $h^\beta = h^{\beta'} \cdot h_l$, where $|\beta'| = n - 1$. Then we obtain a chain of equalities

$$f^\alpha \cdot h^\beta = f_k \cdot h^{\beta'} h_l = (h^{\beta'} f_k + T_0^\alpha(h^{\beta'}; h)) \cdot h_l$$
$$= h^{\beta'} h_l f_k + h^{\beta'} [f_k, h_l] + T_0^\alpha(h^{\beta'}; h) h_l = h^\beta f^\alpha + T_0^\alpha(h^\beta; h) f^0,$$

which shows that the desired polynomials exist for $|\alpha| = 1$ and every β.

If $|\alpha| > 1$, then $f^\alpha = f^{\alpha'} \cdot f^{\alpha''}$, where $\max \mathrm{supp}(\alpha') \leq \min \mathrm{supp}(\alpha'')$ and $0 < \alpha' < \alpha$. Using the induction hypothesis we obtain

$$f^\alpha \cdot h^\beta = f^{\alpha'} f^{\alpha''} \cdot h^\beta = f^{\alpha'} \left(h^\beta f^{\alpha''} + \sum_{\gamma < \alpha''} T_\gamma^{\alpha''}(h^\beta; h) f^\gamma \right)$$
$$= \left(h^\beta f^{\alpha'} + \sum_{\delta < \alpha'} T_\delta^{\alpha'}(h^\beta; h) f^\delta \right) f^{\alpha''}$$
$$+ \sum_{\gamma < \alpha''} \left(T_\gamma^{\alpha''}(h^\beta; h) f^{\alpha'} + \sum_\delta T_\delta^{\alpha'}(T_\gamma^{\alpha''}; h) f^\delta \right) f^\gamma$$
$$= h^\beta f^\alpha + \sum_{\gamma < \alpha} T_\gamma^\alpha(h^\beta; h) f^\gamma.$$

The uniqueness of the polynomials $T_\gamma^\alpha(P; h)$ follows easily from the Birkhoff-Witt theorem. Lemma 3.1 is proved.

Now we return to the proof of sufficiency in Proposition 3.1. Suppose that a Lie algebra $\mathfrak{L}$ satisfies condition $(*)$ and that there exist square matrices P and X over $U(\mathfrak{L})$ such that

$$X(1 + P) = 0 \qquad (3)$$

(the case when $(1 + P)X = 0$ is considered analogously). Write out the matrices P and X over the base (1) of the algebra $U(\mathfrak{L})$ and denote by $\mathfrak{L}(P)$ the subalgebra of $\mathfrak{L}$ generated by the set of those elements from $\mathscr{H} \cup \mathscr{F}$ which occur in that representation of the matrix P. The algebra $\mathfrak{L}(X)$ is defined analogously. It follows from the Birkhoff-Witt theorem that it is sufficient to consider the case $\mathfrak{L}(X) \subseteq \mathfrak{L}(P)$; in particular, we can assume that the equality $\mathfrak{L}(P) = \mathfrak{L}$ holds. Now we carry out induction on the number $\deg X$

In the case when $\deg X \leq 1$ choose the matrix P of the least degree satisfying the relation (3) and write the matrices P and X in the form (2): $P = \sum_\beta P_\beta(h) f^\beta$, $X = \sum_j X_j h_j + \sum_k Y_k f_k$, where X_j ($j \in J$) and Y_k ($k \in K$) are matrices over the ground field F. Assume, on the contrary, that $X \neq 0$. Factoring the equality (3) modulo the ideal generated by the set $\mathscr{H}$ and taking into consideration $U(\mathfrak{L})/ug(\mathscr{H}) \cong U(\mathfrak{L}/[\mathfrak{L}, \mathfrak{L}])$ we obtain the relations $Y_k = 0$ for all $k \in K$. Equating the coefficients of f^β in (3) to zero, we arrive, in particular, at the equality $X(1 + P_0(h)) = 0$ of the same form as before. Here, by the condition $(*)$, $P \neq P_0(h)$. Therefore, the number of monomials in the expression (2) for the matrix $P_0(h)$ is smaller than that in the expression for the matrix P. Turning from the algebra $\mathfrak{L}$ to the algebra $\mathfrak{L}(P_0(h))$ and repeating the above argument several times we finally decrease the degree of the matrix P in the relation (3). The contradiction obtained completes our consideration of the case $\deg X \leq 1$.

Now we make the induction step. To this end we write the matrices P and X in the form (2) again. Multiplying through in (3) by Lemma 3.2, we arrive at the equality

$$\sum_\delta \left\{ X_\delta(h) + \sum_{\{\gamma,\beta|\gamma+\beta=\delta\}} X_\gamma(h)P_\beta(h) \right.$$
$$+ \sum_{\{\gamma,\beta|\gamma+\beta>\delta\}} X_\gamma(h)P_\beta(h)S_\delta^{\gamma\beta}(h) + \sum_{\{\gamma,\beta,\xi|\xi<\gamma,\xi+\beta=\delta\}} X_\gamma(h)T_\xi^\gamma(P_\beta;h) \quad (4)$$
$$\left. + \sum_{\{\gamma,\beta,\xi|\xi<\gamma,\xi+\beta>\delta\}} X_\gamma(h)T_\xi^\gamma(P_\beta;h)S_\delta^{\xi\beta}(h) \right\} f^\delta = 0.$$

Equating the coefficients of f^δ to zero and collecting similar terms at $X_\gamma(h)$ in (4) we obtain a system of equations which implies, in particular, the matrix relation

$$X^{(1)}(h)(1 + P^{(1)}(h)) = 0, \quad (5)$$

where $X^{(1)}(h)$ and $P^{(1)}(h)$ are square block matrices whose blocks are determined by the equality (4) in a unique way (for example, $X^{(1)}_{\gamma\delta} = X_\delta(h)$, $0 \leq \gamma$, $\delta \leq \sup\{\beta \mid X_\beta(h) \neq 0\}$).

Generally speaking, the blocks $P^{(1)}_{\gamma\delta}(h)$ of the matrix $P^{(1)}(h)$ may have nonzero constant terms. However, this is possible only if $\gamma < \delta$. Therefore, performing, if necessary, elementary transformations of rows in (5), we can assume that $P^{(1)}(h)$ is a matrix over $U(\mathfrak{L})$. Also, it is clear that $X^{(1)}(h) \neq 0$ for $X \neq 0$ and $\deg X^{(1)}(h) \leq \deg X$.

As before, we turn from the algebra $\mathfrak{L}$ to the algebras $\mathfrak{L}(P^{(1)})$ and repeat the above argument several times. At a certain step we either obtain the inequality $\deg X^{(n)} < \deg X$ which makes it possible to apply the induction hypothesis, or, for sufficiently large $\overline{n} \in \mathbb{N}$, the relations $X^{(n)} = X_0^{(n)}(h)$ hold for all $n \geq \overline{n}$ (here we have in mind the representation of the matrix $X^{(n)}$ in the form (2)). In the latter case $P^{(n+1)}(h) = P_0^{(n)}(h)$ for all $n \geq \overline{n}$ and the number of monomials in the form (2) of the matrix $p^{(\overline{n})}$ decreases at each step. Thus, for certain $n \geq \overline{n}$, we obtain $P^{(n)} = 0$ and $X^{(n)} = 0$. Q.E.D.

Proposition 3.1, and hence, Theorem 3.1 are proved completely.

4. The undecidability of the word problem in the class of radical rings.

Let Θ be an arbitrary finite set of terms of signature (type) τ, and let $\mathfrak{A}$

be a certain class of algebraic systems of the same signature. Suppose that $\mathfrak{A}(\Theta)$ denotes the subclass of the class $\mathfrak{A}$ that consists of all those algebraic systems on which the set Θ of formulas is valid, and let $W(\Theta)$ be the set of all terms of the signature τ of the same variables as Θ that hold on each algebraic system of the class $\mathfrak{A}(\Theta)$ simultaneously with Θ (that is, under the same interpretation of the variables). If the set $W(\Theta)$ is recursive for every Θ, they say that the word problem is decidable in the class $\mathfrak{A}$. Macintyre [38] and [39] proved the undecidability of the word problem for the class of division rings. We obtain an analogous result for the class of radical rings and at the same time give a simpler proof for division rings.

THEOREM 3.2. *The word problem is undecidable in the classes of radical rings and division rings.*

PROOF. Let $\mathfrak{L}$ be a solvable Lie algebra finitely presented in the class of all Lie algebras with undecidable word problem (the existence of such a Lie algebra was proved by Kukin [18]). Construct a set Θ from the defining relations of the algebra $\mathfrak{L}$ rewritten in the signature of rings (that is, all Lie commutators are replaced by associative ones). We claim that the corresponding set $W(\Theta)$ is not recursive. Indeed, let $\mathfrak{L} \cong F_L\langle \mathfrak{X}\rangle/I$, where $F_L\langle \mathfrak{X}\rangle/I$ is a free Lie algebra and I is the ideal generated by the defining relations of the algebra $\mathfrak{L}$. The recursiveness of the set $W(\Theta)$ would imply the recursiveness of the set I, as Lemma 3.3 shows, and that contradicts our choice of the algebra $\mathfrak{L}$.

LEMMA 3.3. *Let $l \in F_\mathfrak{L}\langle \mathfrak{X}\rangle/I$. Let $\hat{l}$ denote the term in the signature of rings obtained from l by replacing all Lie commutators by associative ones. Then $l \in I$ if and only if $\hat{l} \in W(\Theta)$.*

PROOF. It is clear from the definition of the set Θ that, for every radical (division) ring R on which the formulas of Θ hold, there exists a Lie ring homomorphism $\mathfrak{L} \to R^{(-)}$. Therefore, if $l \in I$ then $\hat{l} \in W(\Theta)$. The converse holds because the ring $U(\mathfrak{L})$, satisfying the formulas of Θ, is embeddable in a division ring and in a radical ring (Theorem 3.1).

5. The author is not aware of examples of Lie algebras satisfying the condition (∗) and not representable in radical rings. It seems to be interesting to find an answer to the following problem: is a universal envelope of a Lie algebra with a single defining relation embeddable in a radical ring?

Chapter 4
Rational Identities of Radical Algebras

The class of associative radical algebras over a commutative ring F forms a variety in the signature consisting of the operation of F-algebras and the operation of taking the quasi-inverse element with respect to the adjoined multiplication $a \circ b = a + b + ab$. For an arbitrary set $\mathfrak{X}$ let $U_J\langle\mathfrak{x}\rangle$ denote the free algebra in this variety with the elements of $\mathfrak{X}$ as free generators, and call this algebra the *free radical*.

An element $\mathfrak{p} \in U_J\langle\mathfrak{X}\rangle$ is called a rational identity of a radical algebra R if for every homomorphism $\varphi\colon U_J\langle\mathfrak{X}\rangle \to R$, the equality $\varphi(\mathfrak{p}) = 0$ holds. If $\mathfrak{p} \neq 0$, the rational identity is called *nontrivial*. Let the symbol $I(R)$ denote the set of all rational identities of radical algebra R.

Amitsur [25] was the first to introduce the notion of a rational identity for the class of division rings (the corresponding definition is obtained from the previous one by replacing the algebra $U_J\langle\mathfrak{X}\rangle$ by the universal division ring $U\langle\mathfrak{X}\rangle$ of quotients of the free algebra $F\langle\mathfrak{X}\rangle$, where F is a field, and the homomorphism $\varphi\colon U_J\langle\mathfrak{X}\rangle \to R$ by a specialization of the division ring $U\langle\mathfrak{X}\rangle$ into the division ring R). He also proved that a division ring that satisfies a nontrival rational identity and possesses an infinite center is a PI-ring, that is, a certain polynomial identity holds on it. Later the proof of this result was simplified by Bergman [26] and Cohn [30].

The problem of whether a radical F-algebra satisfying a nontrivial rational identity is a PI-algebra was formulated by V. N. Gerasimov in [15, Problem 2.31]. The goal of the present chapter is the proof of the following theorem:

THEOREM 4.1. *Let R be a radical algebra over an infinite field F and $I(R) \neq 0$. Then R is a PI-algebra.*

The second section of this chapter is dedicated to the proof of this result. The following Theorem 4.2, which is proved in the third section, is the key to the proof.

THEOREM 4.2. *Let R be a prime radical algebra over an arbitrary field F, and assume $I(R) \neq 0$. Then R is an order in a matrix ring D_m over a division ring D.*

The first section is devoted to finding a certain form of an element $\mathfrak{p} \in U_J\langle \mathfrak{X} \rangle$ that facilitates computations. Here we use notions and notations of Chapter I of the dissertation.

§1. Rational identities of a special form

Fix an arbitrary field F which will be the ground field for all algebras considered in this chapter.

It is clear that the free radical algebra $U_J\langle \mathfrak{X} \rangle$ mentioned in the introduction is nothing other than a universal radical $F\langle \mathfrak{X} \rangle$-algebra, and a rational identity of a radical algebra R is an element $\mathfrak{p} = [(\frac{p' \mid \tilde{p}}{1 + P_0 \mid 'p})] \in U_J\langle \mathfrak{X} \rangle$ such that the equality

$$\tilde{p}(r_1, \ldots, r_n) - p'(r_1, \ldots, r_n)(1 + P_0(r_1, \ldots, r_n))^{-1} \, 'p(r_1, \ldots, r_n) = 0$$

holds for all $r_1, \ldots, r_n \in R$. Here and in the sequel the symbol n is used to designate the number of elements of the set $\mathfrak{X}$ that occur in the expression for a fixed representative $(\frac{p' \mid \tilde{p}}{1 + P_0 \mid 'p})$ of the equivalence class $\mathfrak{p}$.

We assume that the set $\mathfrak{X}$ is well ordered. This order is naturally extended to the semigroup with $\mathfrak{X}$ as the set of free generators. On the algebra $F\langle \mathfrak{X} \rangle$ in the usual way a function $v \colon F\langle \mathfrak{X} \rangle \to \mathbb{N} \cup \{+\infty\}$ of a lower degree is defined and extended to matrix rings over $F\langle \mathfrak{X} \rangle$.

A matrix of the form $(\frac{p' \mid \tilde{p}}{1 + P_0 \mid 'p})$ is called *left* (*right*) *homogeneous of degree* k if the matrices $P_0, 'p$ (p') are homogeneous of degree 1 and the matrices $\tilde{p}, p'('p)$ are homogeneous of degree $k - 1$. Sometimes such a matrix will be simply called (k)-*homogeneous*. Homogeneous matrices of the just described structure with the property $v(\tilde{p}) < v(p' \cdot 'p)$ are called *reduced*; moreover, if the polynomial $\tilde{p}$ is multilinear, then the matrix is called *multilinear*. It is clear that, $\tilde{p} \neq 0$ always holds for a reduced matrix. The notions introduced will be used with respect to the element $\mathfrak{p} = [(\frac{p' \mid \tilde{p}}{1 + P_0 \mid 'p})] \in U_J\langle \mathfrak{X} \rangle$ if they are applicable to a matrix singled out as a representative of the coset.

If an element $\mathfrak{p} = [(\frac{p' \mid \tilde{p}}{1 + P_0 \mid 'p})] \in U_J\langle \mathfrak{X} \rangle$ is homogeneous, then the polynomials $\tilde{p}, (-1)^{i+1} p' P_0^i \, 'p \in F\langle \mathfrak{X} \rangle$ ($i = 0, 1, \ldots$) are called its *homogeneous components*. This definition is justified by the fact that the polynomials indicated are homogeneous components of the image of the element $\mathfrak{p}$ under the natural homomorphism $\Theta \colon U_J\langle \mathfrak{X} \rangle \to F\langle\langle \mathfrak{X} \rangle\rangle$ into the algebra of formal power series on the set $\mathfrak{X}$ of variables. Besides, the following lemma holds:

LEMMA (P. M. Cohn). *The homomorphism* $\Theta \colon U_J\langle \mathfrak{X} \rangle \to F\langle\langle \mathfrak{X} \rangle\rangle$ *is an embedding.*

A proof can be found in [32, p. 142].

LEMMA 4.1. *For every coset* $\mathfrak{p} \in U_J\langle \mathfrak{X} \rangle$ *there exists a* 1-*homogeneous representative.*

PROOF. Among all the matrices belonging to the coset $\mathfrak{p}$ choose a matrix $\left(\frac{p' \mid \tilde{p}}{1+P_0 \mid {}'p}\right)$ with the least senior word u. If $v(u) > 1$ then $u = xw$, where $x \in X$, $v(w) \geq 1$. Let $\tilde{p} = \tilde{\pi} \cdot u + \tilde{p}_1$, $p' = \pi' \cdot u + p'_1$, ${}'p = {}'\pi \cdot u + {}'p_1$, $P_0 = \pi_0 \cdot u + P_0^*$, where $\left(\frac{\pi' \mid \tilde{\pi}}{\pi_0 \mid {}'\pi}\right)$ is a square matrix over F and the senior word of the element $\left(\frac{p'_1 \mid \tilde{p}_1}{P_0^* \mid {}'p_1}\right)$ is less than u. Then we have the equality

$$\begin{pmatrix} p'_1 & -\pi'x & -\tilde{\pi}x & -p'_1 \\ 1+P_0^* & -\pi_0 x & -'\pi x & -P_0^* \\ w \cdot 1 & 1 & 0 & -w \cdot 1 \\ 0 & 0 & 1 & 0 \end{pmatrix} \begin{pmatrix} P_0 & {}'p \\ w \cdot 1 & 0 \\ 0 & w \\ 1+P_0 & {}'p \end{pmatrix} = \begin{pmatrix} -p' & -\tilde{\pi}u \\ 0 & {}'p_1 \\ 0 & 0 \\ 0 & w \end{pmatrix},$$

which, together with Corollary 1 to Theorem 1.4, shows that

$$\mathfrak{p} = \left[\left(\frac{p' \mid \tilde{p}}{1+P_0 \mid {}'p}\right)\right] = \left[\begin{pmatrix} p'_1 & -\pi'x & -\tilde{\pi}x & \tilde{p}_1 \\ 1+P_0^* & -\pi_0 x & -'\pi x & {}'p_1 \\ w \cdot 1 & 1 & 0 & 0 \\ 0 & 0 & 1 & w \end{pmatrix}\right],$$

where the senior word of the latter representative is less than u. Therefore, $v(u) = 1$. Lemma 1.1 is proved.

COROLLARY 1. *For every nonzero coset* $\mathfrak{p} \in U_J\langle \mathfrak{X} \rangle$ *there exists a left (right) reduced representative.*

PROOF. Let $\left(\frac{p' \mid \tilde{p}}{1+P_0 \mid {}'p}\right)$ be a 1-homogeneous representative. If $\tilde{p} \neq 0$ then it is the desired one. Otherwise, by Cohn's lemma, there exists a least natural number k such that $p' P_0^k \,{}'p \neq 0$. Then the matrix

$$\left(\frac{(-1)^{k+1} p' P_0^{k+1} \mid (-1)^{k+1} p' P_0^k \,{}'p}{1+P_0 \mid {}'p}\right) \left(\left(\frac{p' \mid (-1)^{k+1} p' P_0^k \,{}'p}{1+P_0 \mid (-1)^{k+1} P_0^{k+1} \,{}'p}\right)\right)$$

can be taken as a left (right) reduced representative. Corollary 1 is proved.

COROLLARY 2. *If a radical algebra R satisfies a nontrivial rational identity, then some left (right) multilinear rational identity holds in it.*

This statement is obtained by applying the usual linearization to the left- (right-) reduced representative of the rational identity.

§2. Proof of Theorem 4.1 by means of Theorem 4.2

In this section Theorem 4.1 will be deduced from Theorem 4.2 to whose proof the next section is dedicated.

First of all, observe several simplest properties of rational identities.

LEMMA 4.2. *Let* R, S, R_k $(k \in K)$ *be radical algebras. Then the following statements hold*:

(a) *If there exists an embedding* $R \to S$, *then* $I(S) \subseteq I(R)$.
(b) *If there exists an epimorphism* $R \to S$, *then* $I(R) \subseteq I(S)$.
(c) $I(\prod_{k \in K} R_k) = \bigcap_{k \in K} I(R_k)$
(d) *Let* $\mathfrak{p} = [(\frac{p'}{1+P_0}\Big|\frac{\tilde{p}}{'p})]$ *be a multilinear element of the algebra* $U_J\langle\mathfrak{X}\rangle$. *An arbitrary mapping* $\Theta: \mathfrak{X} \to R$ *can be extended in a unique way to a homorphism* $\Theta_t: U_J\langle\mathfrak{X}\rangle \to R[[t]]$ *that satisfies the conditions* $\Theta_t(x) = \Theta(x)t$. *If* $\mathfrak{p} \in \bigcap_\Theta \ker\Theta_t$, *then all homogeneous components of the element* $\mathfrak{p}$ *are polynomial identities over* R.

PROOF. The first three statements are obvious, we prove the forth one. Let $v(\tilde{p}) = k$, $v(p' \cdot {}'p) = m$. Then $k < m$ (since the element $\mathfrak{p}$ is multilinear) and, moreover,

$$0 = \Theta_t(\mathfrak{p}) = \tilde{p}(y_1, \ldots, y_n)t^k$$
$$+ \sum_{i=0}^\infty p'(y_1, \ldots, y_n)P_0^i(y_1, \ldots, y_n)'p(y_1, \ldots, y_n)(-1)^i t^{m+i},$$

where $y_j = \Theta(x_j)$ $(1 \leq j \leq n)$. It follows that for every $i \geq 0$

$$\tilde{p}(y_1, \ldots, y_n) = p'(y_1, \ldots, y_n)P_0^i(y_1, \ldots, y_n)'p(y_1, \ldots, y_n) = 0, \quad \text{Q.E.D.}$$

Now we begin the proof of Theorem 4.1, which we formulate in the following more precise form:

THEOREM 4.1'. *If a radical algebra* R *over an infinite field satisfies a multilinear rational identity* $\mathfrak{p} = [(\frac{p'}{1+P_0}\Big|\frac{\tilde{p}}{'p})]$, *then a polynomial identity* $\tilde{p}^N$ *holds on* R *for a certain natural* N; *for a semiprime algebra* R *we can also assume that* $N = 1$.

PROOF. First we consider the case of a prime algebra. Then, by Theorem 4.2, R is an order in a matrix ring D_m of order m over a certain division ring D.

To each element $\alpha \in F$ there corresponds a homomorphism $\varphi_\alpha: R[t] \to R$ which is identical on R and transforms t into α. It can be extended in a unique way up to a homomorphism $\hat{\varphi}_\alpha: U_J(R[t]) \to R$. Let $\hat{\varphi}_\infty$ denote the homomorphism $U_J(R[t]) \to R^F$ determined by the equality $m = \hat{\varphi}_\alpha(\mathfrak{p})\langle\alpha\rangle = \hat{\varphi}_\infty(\mathfrak{p})$, where $p \in U_J(R[t])$, $\alpha \in F$, and let K_∞ denote its kernel. Obviously, $R \hookrightarrow U_J(R[t])/K_\infty \hookrightarrow R^F$. Applying statements (a) and (c) of Lemma 4.2 to these inclusions, we obtain $I(R) = I(U_J(R[t])/K_\infty)$. Because of this equality and statement (d) of Lemma 4.2, to prove the theorem it suffices to construct, for the homomorphism $\Theta_t: U_J\langle\mathfrak{X}\rangle \to R[[t]]$ (see

Lemma 4.2 (d)), a commutative diagram of the following form:

$$
\begin{array}{ccccc}
U_J\langle \mathfrak{X}\rangle & \xrightarrow{\eta_t} & U_J(R[t]) & \xrightarrow{\pi} & U_J(R[t])/K_\infty \\
\| & \nearrow\sigma & \downarrow i & \nearrow \omega & \\
U_J\langle \mathfrak{X}\rangle & \xrightarrow[\Theta_t]{} & R[[t]] & \searrow \hat{i} &
\end{array}
$$

where i is the natural embedding, π the projection, the mappings σ and $\hat{i}$ appear from the definition of the ring $U_J(R[t])$, so that the triangle that corresponds to them is commutative, the homomorphism η_t is uniquely determined by the property $\eta_t(x) = \Theta(x)t$, and for the existence of the mapping ω, it is sufficient to verify that $K_\infty \subseteq \ker \hat{i}$. This inclusion will follow from Lemma 4.3 given below. Indeed, we only need to understand that regular elements of the ring $R[t]$ remain regular under the homomorphism i. The latter statement is true because R is an order in D_m, and hence, every element of $R[t]$ can be considered as a block matrix over $D[t]$. If this matrix is a zero divisor in $R[[t]]$, then it is degenerate as a matrix over $D[[t]]$, and hence, over $D[t]$ as well. Again, using the fact that R is an order in D_m, we obtain that this element is not regular in the ring $R[t]$ either.

LEMMA 4.3. *For every element $\kappa \in K_\infty$ there exists a polynomial $v(t) \in R[t]$ that is regular in $R[t]$ and such that $\kappa v(t) = 0$. In particular, for every homomorphism $f: R[t] \to S$ into a radical ring S that transfers regular elements into regular ones, its uniquely determined extension $\hat{f}: U_J(R[t]) \to S$, whose existence is guaranteed by the universality property of the ring $U_J(R[t])$, contains the set K_∞ in its kernel.*

PROOF. First we verify the statement on the homomorphsims. Let $f: R[t] \to S$ be a homomorphism that preserves regularity, $\kappa \in K_\infty$ and let $v(t)$ be a polynomial regular in $R[t]$ and such that $\kappa \cdot v(t) = 0$. Then $0 = \hat{f}(\kappa \cdot v(t)) = \hat{f}(\kappa) \cdot f(v(t))$. It follows that $\hat{f}(\kappa) = 0$. Q.E.D.

Now let $\kappa = [(\frac{a'(t)}{1+A(t)} | \frac{\tilde{a}(t)}{'a(t)})] \in K_\infty$. Since R is an order in D_m, the matrix chosen as a representative of the element κ can be considered as a matrix over $D_m[t]$, or as a block matrix over $D[t] \subseteq D(t)$. Here the matrix $1 + A(t)$ is obviously regular over $D[t]$. Therefore, there exist a matrix $B_1(t)$ over $D[t]$ and a nonzero element $b_1(t) \in D[t]$ such that $(1 + A(t))B_1(t) = b_1(t) \cdot 1$. Taking into consideration the fact that R is an order in D_m we obtain $(1 + A(t))B(t) = b(t) \cdot 1$, where $B(t)$ is a matrix over $R[t]$, $b(t) \in R[t]$ and $b(t)$ is regular in $D_m[t]$. Therefore, there exist a column vector $'b_1(t)$ over $D_m[t]$ and an element $v_1(t) \in D[t]$ regular in $D_m[t]$ such that $'a(t) \cdot v_1(t) = (b(t) \cdot 1) \cdot {'b_1(t)}$. As before, there exist a column vector $'b(t)$ over $R[t]$ and an element $v(t)$ regular in $R[t]$ that satisfy the relation

$'a(t) \cdot v(t) = (b(t) \cdot 1) \cdot 'b(t)$. Then, for every $\alpha \in F$, we have

$$0 = \hat{\varphi}_\alpha(\kappa \cdot v(t)) = \tilde{a}(\alpha)v(\alpha) - a'(\alpha)(1 + A(\alpha))^{-1'}a(\alpha)v(\alpha)$$
$$= \tilde{a}(\alpha)v(\alpha) - a'(\alpha)B(\alpha)'b(\alpha).$$

We see that the element $\tilde{a}(t)v(t) - a'(t)B(t)'b(t) \in R[t]$ vanishes for every value of $\alpha \in F$. Since the field F is assumed to be infinite, we have $\tilde{a}(t)v(t) = a'(t)B(t)'b(t)$. We claim that $v(t)$ is the desired element. Indeed,

$$\kappa \cdot v(t) = \left[\left(\begin{array}{c|c} a'(t) & \tilde{a}(t)v(t) \\ \hline 1 + A(t) & 'a(t)v(t) \end{array} \right) \right] = 0,$$

since

$$\left(\begin{array}{c|c} a'(t) & -\tilde{a}(t)v(t) \\ \hline 1 + A(t) & -'a(t)v(t) \end{array} \right) \cdot \left(\begin{array}{c|c} B(t)'b(t) & B(t)'b(t) \\ \hline 1 & 0 \end{array} \right) = \left(\begin{array}{c|c} 0 & \tilde{a}(t)v(t) \\ \hline 0 & 'a(t)v(t) \end{array} \right) \quad \text{Q.E.D.}$$

Lemma 4.3 and with it Theorem 4.1 for a prime algebra are proved.

Every semiprime algebra R is a subdirect product of prime ones which, by Lemma 4.2, satisfy the same multilinear rational identity $\mathfrak{p}$ as R. By what we have proved, the polynomial identity $\tilde{p}$ holds in all these prime algebras. Therefore, it holds in R as well.

In the general case we use a known argument due to Amitsur [25]. Consider an algebra $\overline{R} = R^\mathbb{N}/\mathfrak{b}$, where $\mathfrak{B}$ is the Baer radical for $R^\mathbb{N}$. The algebra $\overline{R}$ satisfies a rational identity $\mathfrak{p}$. Besides that it is semiprime, and hence, by what we have proved, the polynomial identity $\tilde{p}$ holds on it. Thus, for any sequences $\xi_i = (r_k^{(i)}, k \in \mathbb{N})$ $(1 \le i \le n)$ of elements of the ring R the following holds: $\tilde{p}(\xi_1, \ldots, \xi_n) = (\tilde{p}(r_k^{(1)}, \ldots, r_k^{(n)}), k \in \mathbb{N}) \in \mathfrak{b}$. Since the Baer radical is a nil ideal, for arbitrary elements $\xi_1, \ldots, \xi_n \in R^\mathbb{N}$ there exists a natural number m such that

$$\tilde{p}(\xi_1, \ldots, \xi_n)^m = (\tilde{p}(r_k^{(1)}, \ldots, r_k^{(n)})^m, k \in \mathbb{N}) = 0.$$

This shows that the set $\{m \in \mathbb{N} \mid \tilde{p}(r_1, \ldots, r_n)^m = 0 \text{ for all } r_1, \ldots, r_n \in R\} \ne \varnothing$. Therefore, the algebra R satisfies the polynomial identity $\tilde{p}^N$ for some number $N \in \mathbb{N}$.

Theorem 4.1 is completely proved.

§3. Proof of Theorem 4.2

It is sufficient to prove that a prime algebra R (over an arbitrary field) satisfying a multilinear rational identity $\mathfrak{p} = [(\frac{p'}{1+P_0} | \frac{\tilde{p}}{'p})]$ is a Goldie ring. (All definitions and statements necessary for understanding this section can be found in the books [23, §§7.2, 7.3], [2, vol. II, §§10, 11], [19, §4.7].)

LEMMA 4.4. *Let R be a prime radical algebra satisfying a multilinear rational identity. Then the maximality and minimality conditions for left and right annihilators hold in R.*

PROOF. Let $J_1 \subset J_2 \subset \cdots$ be an ascending chain of right annihilators in R, and let $J_S = \mathrm{ann}_r(K_S)$ $(S = 1, 2, \ldots)$, where K_S is a left ideal of the ring R. Then $K_1 \supset K_2 \supset \cdots$ is a descending chain of left ideals. It is clear that $K_{S-1} J_S \neq (0)$ for all $s = 1, 2, \ldots$.

Let $\mathfrak{p} = [(\frac{p' \mid \tilde{p}}{1 + P_0 \mid 'p})]$ be a multilinear element of $U_J \langle \mathfrak{X} \rangle$ such that $\mathfrak{p} = 0$ is a rational identity on R. Without loss of generality we can assume that, in the multilinear polynomial $\tilde{p}$, the coefficient of the monomial $x_1 \cdots x_n$ equals one.

Next choose arbitrary elements $y_S \in J_S K_S$ $(1 \leq S \leq n)$. Then we have $y_i y_j \in J_i K_i J_j K_j = (0)$ for $i \geq j$, and hence, the subalgebra of R generated by the elements $y_1, \ldots, y_n$ is nilpotent of degree not exceeding $n + 1$. Therefore,

$$0 = \tilde{p}(y_1, \ldots, y_n) - p'(y_1, \ldots, y_n)(1 + P_0(y_1, \ldots, y_n))^{-1} {}'p(y_1, \ldots, y_n)$$
$$= y_1 \cdots y_n$$

since $n = v(\tilde{p}) < v(p' \cdot {}'p)$.

Thus, we have proved that $J_1 K_1 \cdots J_n K_n = (0)$. Since the ring R is prime we have $K_1 J_2 \cdots K_{n-1} J_n = (0)$. However, $K_i J_{i+1}$ $(1 \leq i \leq n-1)$ are nonzero ideals of the algebra R. The contradiction obtained shows that R satisfies the maximality condition for right annihilators.

The maximality condition for left annihilators is proved analogously. Lemma 4.4 is proved.

LEMMA 4.5. *Let R be a semiprime ring satisfying the maximality and minimality conditions for right annihilators. Then the following statements hold*:

(a) *The right singular ideal $Z_r(R)$ of the ring R equals zero.*

(b) *If aR is an essential right ideal of the ring R, then the element a is regular.*

(c) *If J is an essential right ideal of the ring R, then J contains an element a such that $\mathrm{ann}_r(a) = (0)$.*

The formulated properties of the ring R are well known (see Lemmas 7.2.1, 7.2.5, and their proofs in [23]) and are stated here only for the reader's sake.

Now it is necessary to verify that, in the conditions of Theorem 4.2, there are no infinite direct sums of right ideals in the algebra R (the finiteness of the Goldie dimension). First we prove this claim for ideals of a special form, and next we obtain a contradiction to the assumption of the existence of such an infinite sum in the general case.

LEMMA 4.6. *Let R be an arbitrary radical algebra and $\mathfrak{p} = [(\frac{p' \mid \tilde{p}}{1 + P_0 \mid 'p})]$ a reduced element of $U_J \langle \mathfrak{X} \rangle$. Suppose that there exist elements $y_1, \ldots, y_n \in R$*

such that $\text{ann}_r(y_i) = (0)$ $(1 \leq i \leq n)$ and
$$\widetilde{p}(y_1, \ldots, y_n) - p'(y_1, \ldots, y_n)(1 + P_0(y_1, \ldots, y_n))^{-1}{}'p(y_1, \ldots, y_n) = 0.$$
Then the sum of right ideals $y_1 R + \cdots + y_n R$ cannot be direct.

PROOF. Assume on the contrary that $y_1 R \oplus \cdots \oplus y_n R$ is a direct sum of right ideals. We can assume that the number $v(\widetilde{p})$ is the least possible. Let
$$\widetilde{p}(x_1, \ldots, x_n) = \sum_{i=1}^{n} x_i \widetilde{p}_i(x_1, \ldots, x_n), \; p'(x_1, \ldots, x_n)$$
$$= \sum_{i=1}^{n} x_i p'_i(x_1, \ldots, x_n).$$
Then
$$0 = \sum_{i=1}^{n} y_i [\widetilde{p}_i(y_1, \ldots, y_n)$$
$$- p'_i(y_1, \ldots, y_n)(1 + P_0(y_1, \ldots, y_n))^{-1}{}'p(y_1, \ldots, y_n)].$$
Our assumption and the conditions imposed on the elements $y_1, \ldots, y_n$ show that the equality
$$\widetilde{p}_i(y_1, \ldots, y_n) - p'_i(y_1, \ldots, y_n)(1 + P_0(y_1, \ldots, y_n))^{-1}{}'p(y_1, \ldots, y_n) = 0$$
holds for every $i \in \{1, \ldots, n\}$. Since $\widetilde{p}(x_1, \ldots, x_n) \neq 0$, there exists a number i such that the element $\mathfrak{p}_i = [(\frac{p'_i}{1+P_0} | \frac{\widetilde{p}_i}{'p})]$ satisfies all conditions stated for $\mathfrak{p}$ and $v(\widetilde{p}_i) < v(\widetilde{p})$ contrary to our assumption. Lemma 4.6 is proved.

COROLLARY. *If the conditions of Theorem* 4.2 *hold, then for any elements* $y_1, y_2 \in R$ *such that* $\text{ann}_r(y_1) = \text{ann}_r(y_2) = (0)$ *the inequality* $y_1 R \cap y_2 R \neq (0)$ *holds. In particular, if* R *is a domain, then it satisfies the Ore condition.*

Indeed, if the conclusion of the corollary were incorrect then the sum $y_2 R + y_1 y_2 R + \cdots + y_1^i y_2 R + \cdots$ of the right ideals would be direct. Together with the existence in R of a multilinear rational identity this contradicts Lemma 4.6.

LEMMA 4.7. *Let* R *be a prime radical algebra with a nontrivial rational identity. If a right ideal* yR *is not essential and* $\text{ann}_r(y) = (0)$, *then* R *possesses a nonzero uniform right ideal (that is, an ideal* U *such that for any nonzero elements* $u, v \in U$ *the inequality* $uR \cap vR \neq (0)$ *holds)*.

PROOF. Since the right ideal yR is not essential, there exists a right ideal J such that $J \neq (0)$ and $yR \cap J = (0)$.

If J is a uniform ideal, the proof is completed. Otherwise, the set
$$A = \{z \in j | (z \neq 0) \& (\exists N((N \triangleleft_r R) \& ((0) \neq N \subseteq J) \& (zR \cap N = (0))))\}$$
is not empty. In this set, choose an element z with a minimal right annihi-

lator. Then an arbitrary nonzero right ideal U of the ring R, which is contained in J and intersects the right ideal zR over zero, is uniform. Indeed, if this is not so, then there exist nonzero elements u, $v \in U$ with the property $uR \cap vR = (0)$. Then for every element $r \in R$ the equality $(z+ur)R \cap vR = (0)$ holds, i.e., $z + ur \in A$. Also, $\text{ann}_r(z + ur) \subseteq \text{ann}_r(z) \cap \text{ann}_r(ur)$. By the choice of the element z, we can assume that $\text{ann}_r(ur) \supset \text{ann}_r(z)$ for every $r \in R$. This means that $uR\,\text{ann}_r(z) = (0)$. Since the algebra R is prime, we can conclude that either $u = 0$ or $\text{ann}_r(z) = (0)$. The former possibility contradicts the choice of the element u, and the latter contradicts the Corollary to Lemma 4.6, Lemma 4.7 is proved.

We need the general density theorem of Amitsur (see, for example, [2.1I, p. 62] for a successful application of the just proved lemma. Before it is stated, we recall certain notions.

A right R-module M_R is called *faithful* (*torsion free*) if for every $r \in R$ (and a nonzero element $m \in M_R$) the equality $Mr = (0)$ ($mr = 0$) implies $r = 0$. The symbol $d(M_R)$ denotes the Goldie dimension of the module M_R.

Let S be a right Ore domain, and let V_S be a right S-module. Let $\overline{V}$ denote the following right module over the division ring $\overline{S}$ of quotients of the ring S: $\overline{V} = V \otimes_S \overline{S}$.

Fix a right Ore domain S and a right S-module V_S. A ring $R \subseteq \text{Hom}_{\overline{S}}(\overline{V}, \overline{V})$ is called a *dense ring of linear transformations* if for every $\varphi \in \text{Hom}_{\overline{S}}(\overline{V}, \overline{V})$ and every submodule $W_S \subseteq V_S$ of finite Goldie dimension $d(W_S) = m < \infty$ there exists a basis $w_1, \ldots, w_m \in W$ of the space $\overline{W}$ and elements $a, b \in R$ such that $(\varphi a - b)w_i = 0$, $aw_i \in \sum_{k=1}^{m} w_k S = W_0$ ($1 \le i \le m$), where the element a is regular in W_0 (this means that the equality $aw = 0$, where $w \in W_0$, is possible only in the case $w = 0$).

DENSITY THEOREM (S. Amitsur). *The following conditions are equivalent for a ring R:*

(1) *R is a prime ring with a nonzero uniform right ideal and the right singular ideal of the ring R equals zero.*

(2) *R is a dense ring of linear transformations in $\text{Hom}_{\overline{S}}(\overline{V}, \overline{V})$, where S is a right Ore domain and V_S a torsion-free module, it contains a transformation r of finite rank (i.e., $d(rV_S) < \infty$).*

Now we return to the proof of Theorem 4.2.

Suppose that the right ideal yR is essential for every element $y \in R$ such that $\text{ann}_r(y) = (0)$. Then $d(R_R) < \infty$. Indeed, let $J = J_1 \oplus J_2 \oplus \cdots$ be an infinite direct sum of right ideals in R. Without loss of generality we can assume that the ideal J is essential. By Lemma 4.5, it has an element $y \in J$ with the property $\text{ann}_r(y) = (0)$. Taking into consideration our assumption, we obtain that yR is an essential right ideal which is impossible because the element y belongs to a finite part of the sum of right ideals considered. Thus, in this case R is a Goldie ring.

It remains to consider the possibility of the existence of an element $y \in R$ that satisfies the condition $\mathrm{ann}_r(y) = (0)$ but generates a nonessential right ideal yR. By Lemma 4.7, R contains a nonzero uniform right ideal and the right singular ideal of the algebra R equals zero by Lemma 4.5. Thus, R satisfies condition (1) of the general density theorem.

It follows that R is a dense ring of linear transformations in $\mathrm{Hom}_{\overline{S}}(\overline{V}, \overline{V})$, where S is a right Ore domain and V_S is a torsion-free module. If $d(V_S) < \infty$, then it is easy to show that R is a Goldie algebra contrary to the case we are considering. Therefore, we can assume that $d(V_S) = \infty$.

Fix an arbitrary natural number m and elements $v_1, \ldots, v_m \in V$ independent over S. We set

$$U = \sum_{k=1}^{m} v_k S \quad \text{and} \quad T = \{r \in R \mid r\overline{U} \subseteq \overline{U}\}$$

and prove that T is a subalgebra of R.

It is relevant to observe here that, although Amitsur's density theorem was stated for rings, it remains valid for algebras. In the case of algebras over a field F we can assume that $F \subseteq S$. Therefore, it remains to verify that T is a subring of R.

Take arbitrary elements $t_1, t_2 \in T$. Then $(t_1 + t_2)\overline{U} \subseteq t_1\overline{U} + t_2\overline{U} \subseteq \overline{U}$, $(t_1, t_2)\overline{U} = t_1(t_2\overline{U}) \subseteq t_1\overline{U} \subseteq \overline{U}$. Therefore, $t_1 + t_2, t_1 t_2 \in T$.   Q.E.D.

Moreover, T is a radical subalgebra of R. Indeed, if t^* is a quasi-inverse for an element $t \in T$ in the ring R, then $t^* \cdot (1+t)\overline{U} \subseteq t\overline{U} \subseteq U$. If the subspace $(1+t)\overline{U} \subseteq \overline{U}$ were proper, then (by the finite dimensionality of U) there would exist a nonzero element $u \in \overline{U}$ such that $(1+t)u = 0$. This contradicts the invertibility of the operator $1+t$ in $\mathrm{Hom}_{\overline{S}}(\overline{V}, \overline{V})$. It follows that $(1+t)\overline{U} = \overline{U}$ and $t^* \in T$.

Let $K = \{r \in R \mid r\overline{U} = (0)\}$. It is clear that K is a two-sided ideal of the algebra T.

We consider the radical ring $\widehat{T} = T/K$ and show that it is an order in the ring of matrices of size m over the division ring $\overline{S}$.

Define the action of an element $\hat{t} = t + K \in \widehat{T}$ onto an element $\overline{u} \in \overline{U}$ in a natural way: $(t + K)\overline{u} = t\overline{u}$. It is seen from the construction of the rings T and K that the result of this action is correctly defined and belongs to $\overline{U}$, while the homomorphism $\widehat{T} \to \mathrm{Hom}_{\overline{S}}(\overline{U}, \overline{U})$ that appears in this way is injective.

Now let $\varphi_0 \in \mathrm{Hom}_{\overline{S}}(\overline{U}, \overline{U}) = \overline{S}_m$. Complete the basis $v_1, \ldots, v_m$ of the space $\overline{U}$ to a basis of the space $\overline{V}$ by elements $v_{m+1}, \ldots$, and extend φ_0 to a transformation $\varphi \colon \overline{V} \to \overline{V}$ by setting

$$\varphi(v_i) = \begin{cases} \varphi_0(v_i), & i \leq m, \\ 0, & i > m. \end{cases}$$

The density of the algebra R in $\mathrm{Hom}_{\overline{S}}(\overline{V}, \overline{V})$ guarantees, for the transformation φ and the finite-dimensional moduie U_S, the existence of a basis $u_1, \ldots, u_m \in U$ of the space $\overline{U}$ and elements $a, b \in R$ such that $(\varphi a - b)u_i = 0$, $au_i \in \sum_{k=1}^{m} u_k S = U_0$ $(1 \leq i \leq m)$ where the element a is regular in U_0 (and hence in $\overline{U}$). It is clear that $a, b \in T$, $(\varphi_0 \hat{a} - \hat{b})u_i = 0$ $(1 \leq i \leq m)$ and that the element $\hat{a} \in \hat{T}$ is regular in $\overline{U}$. It follows that $\hat{a}$ is a regular element of the ring $\hat{T}$ and $\overline{S}_m$ is a classical quotient ring for $\hat{T}$. Q.E.D.

Recall now that R satisfies a nontrivial rational identity. Using Lemma 4.2 we obtain $I(R) \subseteq I(T) \subseteq I(\hat{T})$. If we show that an arbitrarily chosen multilinear element $\mathfrak{p} \in U_J\langle \mathfrak{X} \rangle$ cannot be a rational identity in radical orders of matrix rings D_m over a division ring D for a sufficiently large $m \in \mathbb{N}$, then we shall obtain a contradiction and the proof of Theorem 4.2 will be completed.

LEMMA 4.8. *Let* $\mathfrak{p} = \left[\left(\dfrac{p'}{1+P_0} \bigg| \dfrac{\tilde{p}}{'p} \right) \right]$ *be a multilinear element of* $U_J\langle \mathfrak{X} \rangle$ *and let* $v(\tilde{p}) = n < 2m$. *If* R *is a radical order in* D_m, *then* $\mathfrak{p} \notin I(R)$.

PROOF. By the theorem of Faith-Utumi ([19, p. 183]) we can choose a system of matrix units e_{ij} $(1 \leq i, j \leq m)$ in D_m such that $\sum_{i,j} e_{ij} S \subseteq R \subseteq D_m \subseteq \sum_{i,j} e_{ij} \Delta$, where Δ is the centralizer (which is, of course, a field) in D_m of the set $\{e_{ij} | 1 \leq i, j \leq m\}$, and S is an order in Δ. Analyzing the proof of this theorem (in the above-mentioned book), we see that it is not difficult to understand that S can be considered a radical algebra.

Indeed, $S = \{\sum_{k=1}^{m} e_{k1} u e_{1k} | u \in P \cdot Q\}$, where

$$\mathscr{P} = \{r \in R \mid e_{ij} r \in R \; (1 \leq i, j \leq m)\}, \quad \mathscr{Q} = \{r \in R \mid r e_{ij} \in R \; (1 \leq i, j \leq m)\}$$

(see [19, p. 183]). Therefore, it suffices to verify that an element of the form $a = \sum_{k=1}^{m} e_{k1} u e_{1k}$, where $u \in P \cdot Q$, is right quasi-invertible in the ring S. Set $a^* = -\sum_{k=1}^{m} e_{k1} u(1 - e_{11} u)^{-1} e_{1k}$. Then

$$a \circ a^* = -\sum_{k=1}^{m} e_{k1} u e_{1k} - \sum_{k=1}^{m} e_{k1} u(1 - e_{11} u)^{-1} e_{1k} - \sum_{k=1}^{m} e_{k1} u e_{11} u(1 - e_{11} u)^{-1} e_1$$

$$= \sum_{k=1}^{m} e_{k1} u(1 - (1 - e_{11} u)^{-1}) e_{1k} - \sum_{k=1}^{m} e_{k1} u e_{11} u(1 - e_{11} u)^{-1} e_{1k} = 0.$$

Besides that, if $u = p \cdot q$, where $p \in P$, $q \in Q$, then $u(1 - e_{11} u)^{-1} = pq(1 - e_{11} pq)^{-1} = p \cdot (1 - q e_{11} p)^{-1} q \in P \cdot Q$, since $R^{\#} Q \subseteq Q$. Thus, S is a radical order in the division ring Δ.

Contrary to what we are proving, assume that $\mathfrak{p} \in I(R)$. Choose an arbitrary nonzero element $s \in S$ and consider the value of $\mathfrak{p}$ on the elements $y_1 = e_{11} s$, $y_2 = e_{12} s$, $y_3 = e_{22} s$, $y_4 = e_{23} s$, $\ldots$ assuming that the coefficient of the monomial $x_1 \cdots x_n$ in the polynomial $\tilde{p}$ is equal to zero. We obtain

$$\tilde{p}(y_1, \ldots, y_n) - p'(y_1, \ldots, y_n)(1 + P_0(y_1, \ldots, y_n))^{-1'} p(y_1, \ldots, y_n) = 0.$$

This equality holds in the matrix algebra of order m over the ring $F(s)$, and hence, $p'(y_1, \ldots, y_n) = s^l b'$, $'p(y_1, \ldots, y_n) =' bs$, $(1 + P_0(y_1, \ldots, y_n))^{-1} = (1+B(s))(1+b(s))^{-1}$ for certain vectors $b', 'b$ over F, a matrix $B(s)$ over $F[s]$ and an element $b(s) \in F[s]$. In addition $\tilde{p}(y_1, \ldots, y_n) = s^n e_{11} e_{12} e_{22} e_{23} \cdots$, and this expression does not vanish because of the condition $n < 2m$. Thus, the equality $s^n(1 + b(s)) e_{11} e_{12} e_{22} e_{23} \cdots = s^{l+1} b'(1 + B(s))' b$ holds in the ring $F[s]$ and $n < l + 1$. The element s turns out to be algebraic, and hence, nilpotent (since S is a radical algebra). This contradiction proves Lemma 4.8.

Theorem 4.2 is completely proved.

In connection with Theorem 4.1 many problems arise concerning connections between rational identities of radical algebras and their polynomial identities. We mention only two of them:

(1) Is every homogeneous component of a rational identity of a radical algebra R over an infinite ring its polynomial identity?

(2) Is it true that the ideal $I(R)$ in the free radical algebra $U_J\langle \mathfrak{X} \rangle$ is generated (as a T-ideal) by the set $I(R) \cap F\langle \mathfrak{X} \rangle$?

References

1. V. A. Andrunakievich, *Semiradical rings*, Izv. Akad. Nauk SSSR Ser. Mat. **12** (1948), no. 2, 129–178. (Russian)
2. L. A. Bokut', *Associative Rings*, Parts I and II, Novosibirsk. Gos. Univ., Novosibirsk, 1977 and 1981. (Russian)
3. _____, *Factorization theorems for certain classes of rings without divisors of zero.* I, II, Algebra i Logika **4** (1965), no. 4, 25–52; **4** (1965), no. 5, 17–46; English transl. in Algebra and Logic **4** (1965).
4. _____, *Imbedding of rings into skew fields*, Dokl. Akad. Nauk SSSR **175** (1967), 755–758; English transl. in Soviet Math. Dokl. **8** (1967).
5. _____, *Groups of fractions of multiplicative semigroups of certain rings.* I, II, III, Sibirsk. Mat. Zh. **10** (1969), no. 2, 246–286; **10** (1969), no. 4, 744–799; 800–819; English transl. in Siberian Math. J. **10** (1969).
6. _____, *The problem of Mal'tsev*, Sibirsk. Mat. Zh. **10** (1969), no. 6, 965–1005; English transl. in Siberian Math. J. **10** (1969).
7. A. I. Valitskas, *Absence of a finite basis of quasi-identities for a quasi-variety of rings that can be imbedded in radical rings*, Algebra i Logika **21** (1982), no. 1, 13–36; English transl. in Algebra and Logic **21** (1982).
8. _____, *A representation of finite-dimensional Lie algebras in radical rings*, Dokl. Akad. Nauk SSSR **279** (1984), 1297–1300; English transl. in Soviet Math. Dokl. **30** (1984).
9. _____, *Rational identities of radical algebras*, Izv. Vyssh. Uchebn. Zaved. Mat. **1985**, no. 11, 63–72; English transl. in Soviet Math. (Iz. VUZ) **29** (1985).
10. _____, *Examples of non-invertible rings imbedded into groups*, Sibirsk. Mat. Zh. **28** (1987), no. 2, 35–49, 223; English transl. in Siberian Math. J. **28** (1987).
11. V. N. Gerasimov, *Rings that are nearly free*, Rings, II, Sibirsk. Otdel. Akad. Nauk SSSR, Inst. Mat., Novosibirsk, 1973, pp. 9–19. (Russian)
12. _____, *Inverting homomorphisms of rings*, Algebra i Logika **18** (1979), no. 6, 648–663; English transl. in Algebra and Logic **18** (1979).
13. _____, *Localizations in associative rings*, Sibirsk. Mat. Zh. **23** (1982), no. 6, 36–54; English transl. in Siberian Math. J. **23** (1982).
14. N. Jacobson, *Lie Algebras*, Interscience, New York, 1962.

15. V. A. Andrunakievich (ed.), *The Dniester notebook*, Problems: Theory of Rings and Modules, Akad. Nauk SSSR Sibirsk. Otdel., Inst. Mat., Novosibirsk, 1982. (Russian)
16. P. M. Cohn, *Free Rings and Their Relations*, Academic Press, London and New York, 1971.
17. E. G. Koshevoĭ, *The serving subalgebras of free associative algebras*, Algebra i Logika **10** (1971), no. 2, 183–187; English transl. in Algebra and Logic **10** (1971).
18. G. P. Kukin, *The word problem and free products of Lie algebras and associative algebras*, Sibirsk. Mat. Zh. **24** (1983), no. 2, 85–96; English transl. in Siberian Math. J. **24** (1983).
19. J. Lambek, *Lectures on Rings and Modules*, Blaisdell, Waltham, MA, 1966.
20. A. I. Mal′tsev, *On embedding associative systems in groups*. I, II, Mat. Sb. **6** (1939), no. 2, 331–336; **8** (1940), no. 2, 251–264. (Russian)
21. _____, *Selected Works*. Vol. I. Classical Algebra, "Nauka", Moscow, 1976. (Russian)
22. E. A. Sumenkov, *Certain problems of the theory of Lie algebras*, Diploma Work, Novosibirsk. Gos. Univ., Novosibirsk, 1976. (Russian)
23. I. N. Herstein, *Noncommutative Rings*, Wiley, New York, 1968.
24. S. A. Amitsur, *Rational identities and applications of algebra and geometry*, J. Algebra **3** (1966), 304–359.
25. _____, *Rings with involution*, Israel J. Math. **6** (1968), 99–106.
26. G. M. Bergman, *Skew fields of noncommutative rational functions, after Amitsur*, Séminaire M. P. Schützenberger, A. Lentin et M. Nivat, 1969/70: Problèmes Mathématiques de la Théorie des Automates, Exp. 16, Secrétariat Mathématique, Paris, 1970.
27. A. J. Bowtell, *On a question of Mal′tsev*, J. Algebra **7** (1967), 126–139.
28. P. M. Cohn, *On the embedding of rings in skew fields*, Proc. London Math. Soc. (3) **11** (1961), 511–530.
29. _____, *Dependence in rings*. II. *The dependence number*, Trans. Amer. Math. Soc. **135** (1969), 267–279.
30. _____, *Generalized rational identities*, Ring Theory (Proc. Conf., Park City, UT, 1971), Academic Press, New York, 1972, pp. 107–115.
31. _____, *The embedding of radical rings in simple radical rings*, Bull. London Math. Soc. **3** (1971), 185–188; Correction, **4** (1972), 54; 2nd Correction, **5** (1973), 322.
32. _____, *Free radical rings*, Rings, Modules and Radicals (Proc. Colloq., Keszthely, 1971), North-Holland, Amsterdam, 1973, pp. 135–145.
33. _____, *Inversive localization in Noetherian rings*, Comm. Pure Appl. Math. **26** (1973), 679–691.
34. _____, *The class of rings embeddable in skew fields*, Bull. London Math. Soc. **6** (1974), 147–148.
35. _____, *Skew-Field Constructions*, Cambridge Univ. Press, Cambridge, 1977.
36. R. Doss, *Sur l'immersion d'un semi-groupe dans un groupe*, Bull. Sci. Math. (2) **72** (1948), 139–150.
37. A. A. Klein, *Rings nonembeddable in fields with multiplicative semi-groups embeddable in groups*, J. Algebra **7** (1967), 100–125.
38. A. Macintyre, *The work problem for division rings*, J. Symbolic Logic **38** (1973), 428–436.
39. _____, *Combinatorial problems for skew fields*. I, Proc. London. Math. Soc. (3) **39** (1979), 211–236.
40. P. Malcolmson, *Construction of universal matrix localizations*, Lecture Notes in Math., vol. 951, Springer-Verlag, Berlin and New York, 1982, pp. 117–131.
41. _____, *Matrix localization of n-firs*. I–II, Trans. Amer. Math. Soc. **282** (1984), 503–527.

Translated by BORIS M. SCHEIN

Recent Titles in This Series

(Continued from the front of this publication)

117 **S. V. Bočkarev, et al.,** Eight Lectures Delivered at the International Congress of Mathematicians in Helsinki, 1978
116 **A. G. Kušnirenko, A. B. Katok, and V. M. Alekseev,** Three Papers on Dynamical Systems
115 **I. S. Belov, et al.,** Twelve Papers in Analysis
114 **M. Š. Birman and M. Z. Solomjak,** Quantitative Analysis in Sobolev Imbedding Theorems and Applications to Spectral Theory
113 **A. F. Lavrik,** Twelve Papers in Logic and Algebra
112 **D. A. Gudkov and G. A. Utkin,** Nine Papers on Hilbert's 16th Problem
111 **V. M. Adamjan, et al.,** Nine Papers on Analysis
110 **M. S. Budjanu, et al.,** Nine Papers on Analysis
109 **D. V. Anosov, et al.,** Twenty Lectures Delivered at the International Congress of Mathematicians in Vancouver, 1974
108 **Ja. L. Geronimus and Gábor Szegő,** Two Papers on Special Functions
107 **A. P. Mišina and L. A. Skornjakov,** Abelian Groups and Modules
106 **M. Ja. Antonovskiĭ, V. G. Boltjanskiĭ, and T. A. Sarymsakov,** Topological Semifields and Their Applications to General Topology
105 **R. A. Aleksandrjan, et al.,** Partial Differential Equations, Proceedings of a Symposium Dedicated to Academician S. L. Sobolev
104 **L. V. Ahlfors, et al.,** Some Problems on Mathematics and Mechanics, On the Occasion of the Seventieth Birthday of Academician M. A. Lavrent'ev
103 **M. S. Brodskiĭ, et al.,** Nine Papers in Analysis
102 **M. S. Budjanu, et al.,** Ten Papers in Analysis
101 **B. M. Levitan, V. A. Marčenko, and B. L. Roždestvenskiĭ,** Six Papers in Analysis
100 **G. S. Ceĭtin, et al.,** Fourteen Papers on Logic, Geometry, Topology and Algebra
99 **G. S. Ceĭtin, et al.,** Five Papers on Logic and Foundations
98 **G. S. Ceĭtin, et al.,** Five Papers on Logic and Foundations
97 **B. M. Budak, et al.,** Eleven Papers on Logic, Algebra, Analysis and Topology
96 **N. D. Filippov, et al.,** Ten Papers on Algebra and Functional Analysis
95 **V. M. Adamjan, et al.,** Eleven Papers in Analysis
94 **V. A. Baranskiĭ, et al.,** Sixteen Papers on Logic and Algebra
93 **Ju. M. Berezanskiĭ, et al.,** Nine Papers on Functional Analysis
92 **A. M. Ančikov, et al.,** Seventeen Papers on Topology and Differential Geometry
91 **L. I. Barklon, et al.,** Eighteen Papers on Analysis and Quantum Mechanics
90 **Z. S. Agranovič, et al.,** Thirteen Papers on Functional Analysis
89 **V. M. Alekseev, et al.,** Thirteen Papers on Differential Equations
88 **I. I. Eremin, et al.,** Twelve Papers on Real and Complex Function Theory
87 **M. A. Aĭzerman, et al.,** Sixteen Papers on Differential and Difference Equations, Functional Analysis, Games and Control
86 **N. I. Ahiezer, et al.,** Fifteen Papers on Real and Complex Functions, Series, Differential and Integral Equations
85 **V. T. Fomenko, et al.,** Twelve Papers on Functional Analysis and Geometry
84 **S. N. Černikov, et al.,** Twelve Papers on Algebra, Algebraic Geometry and Topology
83 **I. S. Aršon, et al.,** Eighteen Papers on Logic and Theory of Functions
82 **A. P. Birjukov, et al.,** Sixteen Papers on Number Theory and Algebra

(See the AMS catalog for earlier titles)